Transnational Chinese Cinema:
Corporeality, Desire, and the Ethics of Failure

Transnational Chinese Cinema:
Corporeality, Desire, and the Ethics of Failure
Edited by Brian Bergen-Aurand, Mary Mazzilli and Hee Wai-Siam
Copyright © 2014

Picture 3.1 Still from Love Without End 不了情
© Celestial Pictures Ltd. All rights reserved 版權由天映娛樂有限公司全部擁有

Distributed by Transaction Publishers
10 Corporate Place South, Suite 102
Piscataway, NJ 08854

All rights reserved. Exclusive English language rights are licensed to Bridge21 Publications, LLC. No part of this book may be used or reproduced in any matter whatsoever without written permission from the publisher except in the case of brief quotations embodied in critical articles and reviews.

For information contact Bridge21 Publications, LLC, 11111 Santa Monica Blvd, Suite 220, Los Angeles, CA 90025.

Published in the United States
Cover Design by Chi-Wai Li
Copyedited by Lauren Manoy
ISBN 978-1-62643-010-5 Paperback

Transnational Chinese Cinema
Corporeality, Desire, and the Ethics of Failure

Edited by
Brian Bergen-Aurand
Mary Mazzilli
Hee Wai-Siam

Contents

Preface 7

Introduction 9

CHAPTER 1:
The Ruined Bodies of Transnational Chinese Cinema
Brian Bergen-Aurand 27

CHAPTER 2:
Transnational Cinema as a Matter of Address:
Considering Eros and Embodiment in Wong Kar-wai's "The Hand"
Sim Jiaying 51

CHAPTER 3:
Female Chinese Stars on Screen:
Desiring the Bodies of Ruan Lingyu and Linda Lin Dai
Mary Mazzilli 69

CHAPTER 4:
Trans-mothering on Singapore's Siniticate Screens
Jun Zubillaga-Pow 95

CHAPTER 5:
Coming Out in the Mirror: Rethinking Corporeality and Auteur Theory with Regard to the Films of Tsai Ming-liang
Hee Wai-Siam 113

CHAPTER 6:
Thinking the Inutility: Temporality, Affect, and Embodiment in *Useless* and *Walker*
Hongfei Liao 137

CHAPTER 7:
"Disposable" Bodies on Screen in Xu Xin's *Karamay*: Biopolitics, Affect, and Ritual in Chinese Central Asia
Darren Byler 159

CHAPTER 8:
The Art of Eating in Malaysian Cinema: The Malaysian Sinophone Hunger for a National Identity
Lee Yuen Beng 181

CHAPTER 9:
Random Acts of Sensible Violence: Horror, Hong Kong Censorship, and the Brief Ascent of "Category III"
Andrew Grossman 201

CHAPTER 10:
Drifting Eyeballs: Trans-Asian Feminine Porn Tastes and Experiences
Katrien Jacobs 225

Bibliography 247
Filmography 263
Contributors 269
Index 273

Preface

The essays in this collection focus on transnational Chinese cinema, corporeality, desire, and the ethics of failure. They explore the corporal, psychological, and affective aspects of encountering bodies on screen and engage with the material and discursive elements of embodied moving image experiences to highlight the mind-body dynamics involved in bio-cultural practices of cinematic production, distribution, exhibition, and reception. In approaching these films and videos through embodiment studies, this volume complicates and develops the scope of existing approaches to Chinese cinema, including national, transnational, Chinese-language, and Sinophone models. The writers in this collection draw from a variety of methodologies—including textual and reception analysis, cultural studies, queer theory and feminism, phenomenology, ethics, trauma and diaspora studies, affect studies, questions of film authorship and film stars, genre studies, and debates over censorship—as they contend with incarnate and disincarnate aspects of cinematic and post-cinematic encounters. Engaging with the bio-cultural regimens involved in the different aspects of experiencing Chinese and other bodies entangled in diverse moving image situations allows the authors to explore the relation between the cinematic, the carnal, and the social in regard to cognition, sensation, and emotion, leading to discussions of the ethics of cinema, specifically the ethics of failure. Throughout this collection failing becomes a strategic or practical exercise with the potential to transgress and transmogrify. The final focus of these analyses is on failure, not as opposed to success, but as critical practice stemming from and developing further discussions of ruins and ruination, inutility, touching and drifting, desire and disgust, disillusionment, voicelessness, sacrifice, wounds and scars, arousal, disposability, and hunger. Here, the cinema provokes us to see the possibility of failure.

This book begins with a debt of gratitude to the students enrolled in the Master of Arts in Contemporary China program at Nanyang Technological University, Singapore, in 2013. Valerie Chia, Eunice Lim, William M. Kelly, Isabel Ho Ci Xian, Sung Pik Wan, Ong Leong Kheng Terence, Anthony Liew Kok Keong, Gabriel Lee, Chew Chia Meng, Chia Jia Hao Alvin, Wu Jiayan, Ong Chen Huei, and several others officially and unofficially participated in the CC6390: Special Topics in

Society and Culture—Transnational Chinese Cinemas—Imag(in)ing Embodiment seminar that became the impetus for this book.

In addition to these students, the editors would like to thank the many people who made this project possible, especially Anne, Leake, and Tsegu.

Introduction
Brian Bergen-Aurand

Making conscious sense from our carnal senses is something we do whether we are watching a film, moving about in our daily lives and complex worlds, or even thinking abstractly about the enigmas of moving images, cultural formations, and the meanings and values that inform our existence.
—Vivian Sobchack, *Carnal Thoughts*

Ti is one of the most basic, earliest, and most essential concepts in Chinese philosophy, which derives its meaning from intimate understanding of reality, self, and practice. On its most elementary level, *ti* is the concrete corporeal body that a person possesses, the entity in which human life is maintained and developed. But *ti* is not simply a matter of organization of physical elements. Instead, it is a structure and system of organic functions and vital spirit in the vehicle of the physical body. It may be said that by virtue of the form of the physical body, *ti* realizes its living spirit and vitality, and by virtue of the living spirit and vitality, the physical body maintains its organic unity and organization.
—Chung-Ying Cheng, "On the Metaphysical Significance of *Ti* (Body-Embodiment) in Chinese Philosophy"

This is an entangled book, a collection of essays enmeshed in the networks of transnational Chinese and Sinophone cinema studies, contemporary embodiment studies, and studies of the ethics of failure. Although no single essay in this collection covers all these topics, taken together, they provide an entry into this ensemble of issues, conceptualizations, and articulations. It is surprising, given the centrality and importance of *ti*, or embodiment, to Chinese philosophy, the vast amount of research on Chinese cinema published since the mid-1980s, and the ever-present interest in the connection between embodiment and ethics, that this is the first collection of essays in English to attempt such a sustained encounter. Our goal here is to begin a conversation around questions of corporeality and desire, the mutually constituting material and immaterial aspects of embodiment, and film production, distribution, exhibition, and reception in a transnational Chinese context, and to address this conversation specifically toward the ethics, *phronesis*, or *chresis* involved in encountering transnational Chinese cinema.

It has been over fifteen years since the publication of Sheldon Lu's edited volume *Transnational Chinese Cinemas* (1997) and almost ten years since Lu and Emilie Yeh coedited their collection, *Chinese-Language Film* (2005). In the wake of those two books, much has been written on Chinese cinema around the world, including an ever-increasing amount of scholarship arising within anglophone academia. Countless articles, more than thirty books, and several journals dedicated to Chinese cinema and specific filmmakers, films, and their production and circulation have appeared over the last decade, and the international interest in Chinese cinema seems only to be increasing as scholars focus more intently on specific areas of documentary and alternative genres, women and minority filmmaking and reception, lesbian, gay, bisexual, queer, and trans* filmmakers and filmmaking, and global responses to transnational Chinese film and screen cultures.

Four models have come to set the tone for studying this cinema. The national film model is linked to political and territorial constructs, especially as they are endorsed or contested by the films under consideration. The national film model also takes into account films that share in the national language, geography, and cultural assumptions the nation-state relies on for its persistence. While the national model is most bound to the concerns of film production, the transnational film model opens questions about film distribution, exhibition, and reception as well. Taking into account larger geographies and synergies than the national model, the transnational model crosses and blurs borders, usually to investigate historical or thematic topics, such as comparative studio production analysis, diaspora studies, genre analyses, and issues involving identity politics or considerations of gender and sexuality. The current volume, of course, borrows greatly from this model—not just in its title but also in its examination of embodiment throughout a good number of cinematic locations. The Chinese-Language cinema model eschews questions of the nation-state and political legitimacy, even if it holds to concerns over geography and territoriality. Focused on the languages deployed in the films—predominantly Chinese dialects—this model allows scholars to concentrate more on the film texts themselves and their collaborative (sometimes international) productions than bind them to issues stemming from geopolitical disputes. Such disputes can certainly be taken into consideration within the Chinese-language model, but they need not be the primary barrier to transcultural film studies; nor does the question of "legitimate representation" need to dominate Chinese-language scholars' interpretations. Several of the authors in the present collection borrow from this model as they develop analyses of language and dialect, transnational collaboration, immigration and emigration, and ideological frameworks. The fourth model, as described by Sheldon Lu in his essay, "Notes on Four Major Paradigms in Chinese-Language Film Studies," is the Sinophone cinema approach.[1] As Lu envisions it, the Sinophone model is near to the Chinese-language model but for its further evocation of the colonial, postcolonial, and neocolonial effects made plain in the language, culture, psychology, and quotidian expressions deployed in these films. Intensely engaged with questions of the relations across territory—relations among mainland China and "the peripheries" of the Chinese world—this model is especially helpful in examining co-productions and cross-cultural productions highlighting the relations between centers and margins. As with the other models, several of the studies in the current volume borrow and develop various aspects of the Sinophone model.

To these four models, which Lu explains have "significant points of contrasts as well as overlaps"—and we fully recognize how much our studies intersect with these other four models—we propose an approach to Chinese cinema through the lens of embodiment studies.[2] The embodiment model highlights national, local, and translocal contexts; it draws on international, transnational, and regional relationships; it emphasizes language, culture, history, geopolitical contact zones, and sites of contest—all as they are depicted in relation to the deployment of bodies on screen and at the cinema. Most importantly, since embodiment is always located in time and space, it sees itself as an adjustable approach, a lens with variable focal lengths. It can work with deep focus, in close-up, or it can take a panoramic view. It can crosscut, flash back, flash forward, and compare how bodies signify and affect differently, at different times, in different places. And, these are the two primary sites of interest for the embodiment model—to engage with articulations of desire and corporeality in terms of both signification and affect. How do bodies on screen function, for whom, where, and when?

The starting point for the intersection of screen studies and embodiment studies, of course, is Vivian Sobchack's work in her *The Address of the Eye: A Phenomenology of Film Experience* and her *Carnal Thoughts: Embodiment and Moving Image Culture*.[3] In these two books, Sobchack was able to describe an in-depth analysis of the film experience. More than simply considering what films mean or how we might interpret them as texts, Sobchack opened a conversation into how we see, hear, feel, and react to films—cognitively, emotionally, psychologically, carnally, phenomenologically. Adapting from how Sobchack describes her project, then, the embodied focus here is on the filmic experience and what it is to live in one's body, not merely look at or listen to bodies. We take into account vision, visuality, and visibility, certainly. They remain key aspects for several of the studies here. Yet, we work to treat the visual aspects of film and media as only one element among many. As we watch films, we become excited, bored, intrigued, aroused, disgusted, disappointed, disinterested, and fascinated. Sometimes these reactions come because we experience a different understanding of the cinema; sometimes they come before or after that understanding. Foregrounding our embodied experience of the cinema, what Sobchack describes in *Carnal Thoughts* as our experience of "the lived body as, at once, both an objective subject and a subjective object: a sentient, sensual, and sensible ensemble of materialized capacities and agency that literally and figurally makes sense of, and to, both ourselves and others."[4] Most of all, the studies collected here seek to flesh out our experiences of these cinematic and post-cinematic images.

In this context, then, we pay special attention to the visual, the audio, and the haptic channels of these films as we describe them in detail, following the work of Laura Marks in *The Skin of the Film: Intercultural Cinema, Embodiment, and the Senses* and *Touch: Sensuous Theory and Multisensory Media*.[5] (Careful description of the films is of primary importance to the embodied approach as we try to avoid rushing to judgment and closing down our experiences in interpretations.) The embodiment model calls into question any mind-body duality. We are not disembodied eyes that only gaze at the screen. We are eyes, skin, ears, mouths, hair, emotions, thoughts, memories, yearnings. We are interested in the materiality of the films and the images they deploy. We are interested in the forces and flows that stem from and return to those material deployments. We are interested in the dynamic contact that is the filmic experience. As much as we may know or understand characters, their situations, and their stories, we also feel them and react to them, sometimes as if we are sharing those situations and stories. Consider suspense, surprise, shock, arousal, pain, and pleasure, for example. Consider how we cry or sit on the edges of our seats. Consider how we talk back to the screen or hide our eyes in stressful moments. The embodiment model works to take into account all these aspects of the films it considers.

In its resistance to the mind-body dichotomy, the embodiment model parallels well with Chinese thought and Chinese cinema. As Chung-Ying Cheng explains,

Throughout the history of Chinese philosophy, the principle of substance and function in unity is well maintained and cherished. This explains why in the history of Chinese philosophy there is the absence of the fundamental dualism of mind and body à la Descartes, or the fundamental dualism of reality and appearance à la Plato, or the fundamental dualism of knowledge or understanding of objects and rational intuition of things-in-themselves à la Kant. From the Chinese point of view of the *ti-yong heyi* or *ti-yong buer*, mind and body are both functions of the deeper reality—substance of the human person, which is his nature, his principle of being, or his destiny (*ming*).[6]

Separating the mind and body or representation and reality have never been fundamental in Chinese philosophy. Rather, the fundamental has been the unified "substance of the human person," which makes the divisions of mind and body possible, without necessarily conceptualizing them as the only two divisions possible. They are two divisions among possible others. The chapters in this book explore the relation between that "substance" and its possible divisions as it develops in the cinema by drawing out the complex relation between representation and reality as possible divisions among others with regard to film and screen function.

Considering film carefully, the embodiment model focuses on bodies on screen and bodies before the screen. In doing so, it takes into account both how we understand films and other media and how they affect us. In addressing affect, we rigorously consider film context and film style. We also focus on display, acting, gesture, and movement on screen. Finally, we pay extra attention to how we respond to filmed bodies as corporeal, desirous bodies ourselves. We are especially interested in how our cinematic and post-cinematic experiences alter our experiences of our own and others' bodies. Taking particular note of how filmic elements function to provoke affects and how affect functions to remind us of the body's relation to the self—as constitutive of the self and sometimes foreign to the self—the studies collected here return always to the contextual specificity of the tensions between the cinematic and the embodied, their focus is on the intertwining of the filmic experience. Again, according to Sobchack,

> Embodiment is a radically material condition of human being that necessarily entails both the body and consciousness, objectivity and subjectivity, in an irreducible ensemble. Thus we matter and we mean

through processes and logics of sense-making that owe as much to our carnal experience as they do to our conscious thought.[7]

In this way, we recognize that our conscious thought and our carnality are both put into flux by the very experiences we are trying to explain here. The mind and body that experiences the film is altered by the experience. Describing this process of entanglement is at the heart of much of our work in this book as we attempt to articulate the complex mediation, as Sobchack refers to it, that is our embodied experience always:

> However direct it may seem, our experience is not only always mediated by the lived bodies that we are, but our lived bodies (and our experience of them) is always also mediated and qualified by our engagements with other bodies and things. Thus, our experiences are mediated and qualified not only through the various transformative technologies of perception and expression but also by historical and cultural systems that constrain both the inner limits of our perception and the outer limits of our world.[8]

In a way, then, adapting the embodiment model to the study of cinema and post-cinema is somewhat coming full circle. We experience our bodies as mediated, and we experience mediated embodiment through our bodies. Through this dynamic, then, we encounter the film, our selves, and others around us.

In his essay on embodiment, philosophy, and function, Chung-Ying Cheng stresses that embodiment studies "is not something to be understood in separation from a context of cognitive and moral understanding."[9] In fact, what almost every model of embodiment studies eventually turns toward is questions of morality and ethics. Sobchack writes, "the very nature of our embodied existence 'in the flesh' lays the concrete foundations for a materialist—rather than idealist—understanding of aesthetics and ethics." Her hope, she explains, is that thinking through carnality and desire will help us better engage with "our more general and always social entailments with others."[10] Marks's *The Skin of the Film* engages throughout with the question of the relation between memory and life practices. In their introduction, the editors of *Embodied Selves*, Stella Gonzalez-Arnal, Gill Jagger, and Kathleen Lennon, turn toward discussions of how investigations of "the range of encounters between specifically embodied selves and specifically embodied others is foundational to the development of an embodied ethics."[11] And, in their collection *Embodied Modernities: Corporeality, Representation, and Chinese Cultures*, Fran Martin and Larissa Heinrich point out that their approach to Chinese "body-texts" is a study of the practices and texts that have shaped and been influenced by the idea and the materiality of "Chineseness" in the modern period. They are looking as closely at the practices that shaped (and continue to shape) the Chinese body in the modern period as they are at the bodies themselves. In this way, embodiment studies is linked to questions of ethics, questions of ways of being in the world—alone and with others.

Following in this tradition, the chapters here also all eventually turn toward ethics in one way or another, whether as in responsibility, *phronesis* or practical skills and wisdom, or *chresis* or the art and manner of doing something or how something functions. We all end up considering questions of cinema and embodiment in terms of ethics, and we all end up considering questions of cinema, embodiment, and ethics in terms of failure. If there is a thread that runs through all the essays in this book it is that we all turn toward an analysis of ethical practices in regard to experiences of this cinema and post-cinema and in that analysis conclude by discussing the effects of failure on and in that experience. From our surveys, it is possible to suggest that desire and corporeality in transnational Chinese cinema always circulate through an economy of failure. To be clear, this is not to say that any of these films fail or that transnational Chinese cinema is a failure. (Its ever-growing popularity seems to dismiss any such accusation.) What we are saying, though, is that failure is foundational to this cinematic experience. Yet, the ethics of failure we describe marks failure as potential, possibility, or even promise. This is not to confuse it with hope. Rather, it is to draw a connection between failure and the possibility to continue to practice or practice differently. In this way, a successful practice would be a finite one, one that has been fully conceived, one that has come to fruition. We make no such claim for questions of desire and corporeality in transnational Chinese cinema.

The goal of contemporary body studies is to engage fully with *ti*, with mind and body, to "flesh out" our experiences of daily cultural and social phenomena. As Sobchack writes of her own approach in *Carnal Thoughts*,

> thus, whether exploring how we are oriented spatially both off and on screen or asking about what it means to say that movies "touch us," whether considering the ways in which technology from pens to

computers to prosthetic legs alter the shape of our bodies as well as our lives or the difference between the "visible" and the "visual" in an image-saturated culture, or whether trying to think through the "reality" of certain screen images or the way in which our aesthetic and ethical senses merge and emerge "in the flesh," all the essays in this volume are focused on the lived body. That is, their concern is not merely with the body as an abstracted object belonging always to someone else but also with what it means to be "embodied" and to live our animated and metaphoric existences as the concrete, extroverted, and spirited subjects we all objectively are.[12]

With this summary in mind, we turn to the essays in this volume and what it is they strive to "flesh out." Not all of us adhere to the letter of Sobchack's or any other one methodology or to the strictures of any one phenomenological, materialist, embodiment approach. But we all work, in one way or another, from this view of contemporary embodiment studies.

* * *

In Chapter 1, Brian Bergen-Aurand engages with questions of film ethics regarding the ruined bodies of transnational Chinese cinema. Recent writing on embodiment in transnational Chinese cinema claims that these films no longer project the Chinese body as a reliable marker of authenticity or identity because these films situate "Chineseness" in evermore fragmented, fluid, and ruined relations with their surroundings and link them increasingly with the Chinese diasporic experience. Attempts to essentialize or reify the spectral Chinese body through onscreen spectacle only serves to disperse or erase it further, and calls for a reconceptualization of our practices of depicting Chinese bodies and of maintaining diasporic Chinese viewing positions out of an already impossible fixed, static notion of Chineseness. Rather than further trapping or fixing the image of the Chinese body (and especially the diasporic Chinese body) by making it permanently "visible" and putting it on display, we might rethink the ethics of neighbors and locality in regard to these films and the spectral bodies they imagine. Bergen-Aurand suggests we rethink our ethics toward these "ruined bodies" through the inauthenticity of their screen embodiment. Perhaps, the focus of transnational Chinese cinemas on ruined bodies/bodies in ruins further provokes a reconceptualization of the relation between film ethics and film bodies regarding the ambivalent signification of ruins. While ruins mark collapse, destruction, decay, and bankruptcy, they also mark experience, history, memory, culture, preservation, heritage, and inheritance. Thus, rethinking ruined bodies/bodies in ruins foregrounds a rethinking of the wounds and scars at the origins of diasporic bodies, refocusing diasporic ethics away from origins in times and places toward acts of leaving or being forced to leave: an ethics of bodies relating in transit. This refocus is precisely part of the provocation of such transnational Chinese films as *Enter the Dragon* (1973), *Ju Dou* (1990), *Farewell My Concubine* (1993), *East Palace, West Palace* (1996), *Happy Times* (2000), *In the Mood for Love* (2000), *Be with Me* (2005), *Three Times* (2005), *Spider Lilies* (2007), *Lust, Caution* (2007), and *Uncle Boonmee Who Can Recall His Past Lives* (2010).

Sim Jiaying, in Chapter 2, suggests a new way of thinking the transnational through her analysis of Wong Kar-wai's film "The Hand." Here, she asks that we reconsider the definition of "transnational" in relation to the cinema. Rather than thinking of this organizational concept only in terms of film production, distribution, or exhibition, she suggests we also employ it when considering film reception and style. "Transnational," she argues, refers to a specific affective mode of cinematic address, one we experience in films such as Wong's "The Hand" as the film almost seems to transgress media and space as it "touches us." Sim situates Wong Kar-wai's "The Hand" in its transnational context through its relation to the American and Italian segments of the anthology film *Eros* (2004) and then complicates that situation. The fact that "The Hand" was meant to appeal to an international audience highlights the importance of considering the roles of "modes of spectator address" and reception when considering conceptualizations of transnational cinema. In discussing these conceptualizations, this chapter engages in an examination of the cinematic address and film style of "The Hand" to suggest we reconsider the eros of embodiment and the embodiment of *Eros* in terms of more complex understandings of desire and corporeality. In the end, this reconsideration allows for a rethinking of "transnational cinema" beyond simple categories of geography, signification, and representation.

Also considering the filmic aspects of address, embodiment, and transnationalism, Mary Mazzilli, in Chapter 3, considers two earlier examples of female bodies in transnational Chinese Cinema as she focuses on images of Ruan Lingyu (1910–1934), a female star from the Golden era of Chinese cinema in the 1930s, and Linda Lin Dai (1930–1964), a transnational star of 1950s and 1960s Hong Kong. In discussing representations of female bodies on screen, film scholars have described how the male directorial gaze objectifies the female body and appropriates female subjectivity for its own pleasure. According to such critics, established visual codes and narrative norms subject the female body on screen not only to the desire of the camera gaze but also coerce it into performing and displaying that desire. This self-perpetuating process of manipulation, for some film critics, is not only inevitable but necessary for the production of cinematic pleasure. Against this established paradigm, Mazzilli argues that by looking more closely at the embodiment of particular Chinese female film stars on and off screen, and in particular their images in closeups, we can discover that female stars are often not passive, not victims of the external directorial gaze, but enact their own place in society and their own Nietzschean "will to power," or striving for a superior position. In this chapter, female Chinese stardom is aligned with the confluence of corporeality and desire—in light of Elizabeth Grosz's discussions of embodiment and Merleau-Ponty's conception of visibility—in order to better articulate Chinese female stars' embodied cinematic experience. Thus, it is these female Chinese stars' will to visibility that directly challenges the controlling directorial gaze.

In Chapter 4, Jun Zubillaga-Pow turns toward Singapore's Siniticate screens and the function of trans-mothering in a social context. He contends that the cinematic reproduction of "mothering" is not restrained between the bodies of mothers and children but is shackled to the greater good of Singapore's consumerist society and its precarious nationhood. Even the cross-dressing mother is always and already nationalized and deployed to undermine transnational feminist strategies. Zubillaga-Pow asserts the figure of the Chinese "trans-mother"—a cross-dressing male actor in the role of a mother—is used to delineate how the fictitious role of the mother/widow has remained bound to the familial function of providing parental guidance and infantile protection. In particular, he argues that recent Singapore cinema (and television) has retained the patriarchal tradition of portraying mothers as "homely" as they have been in most Shakespearian and Asian theatre. How these values of vice and virtue come to be embodied in the cross-dressing actor is no more an anti-feminist oddity within Siniticate screen culture than it is in everyday reality. To the extent that these narratives have also shaped cultural perceptions of trans-mothers and their role in bringing up children vis-à-vis family values, cinematic representations have little altered dramatic, filmic, or social attitudes. In the end, Zubillaga-Pow contends that how Singaporeans have come to accept or tolerate gender-changing and cross-dressing as a staple of contemporary visual media is analogous to the constant negotiation of linguistic semantics in the global nation-state, where ethnicities, languages, and beliefs have been disentangled and made homogeneous to alleviate any difference within the national imagination.

Hee Wai-Siam, in Chapter 5, also considers the family and its queer intersection with questions of film authorship in his analysis of the films of Tsai Ming-liang. Rethinking Tsai's oeuvre from the perspective of auteur theory and corporeal narratology, Hee traces the queer lines that run through Tsai's Taiwanese auteur films, TV dramas, and stage productions from the 1980s until today and connects these lines to Tsai's earlier literary works during his time in Kuching, Malaysia. Thus, this essay links Tsai's work in Taiwan to his work in his birthplace, Malaysia, to show the Malaysian Chinese vectors throughout. To demonstrate these links, Hee examines the intersection of auteur theory and corporeality in Tsai Ming-liang's oeuvre in the context of the development of the mirror experience. In doing so, he asks whether Tsai's films create a closet for Chinese queers or construct a queer performativity belonging exclusively to Chinese people. Combining readings of the films and interviews with Xiao Kang and Tsai Ming-liang, this chapter reconceptualizes their mirror relationship and its performative elements from the perspectives of both men. Tsai borrows auteur theory to legitimize the process of creating Xiao Kang's queer body while simultaneously using the creation of the "Xiao Kang household" in the films to found an emotional home for himself as a permanent member of the Malaysian diaspora in Taiwan. So, while it creates this opening, Tsai's self-expression through his films has also always been limited by his mirror relationship with Xiao Kang. Furthermore, this mirror relationship forms a closet that refuses entry to language and understanding, minimizing narrative and interactive logic throughout his works to the point that the majority of audiences must rely on Tsai's extra-diegetic descriptions for their understanding of his films. Hee concludes that Tsai's words, actions, and works have, much of the time, generated a closet for Chinese queers through the mirror experience that reflects a type of queer performativity belonging exclusively to Chinese queers.

Directly addressing the temporal aspects of affect and embodiment, in Chapter 6, Hongfei Liao analyzes two documentary films, *Useless* (Wuyong, 2007) directed by Jia Zhangke (mainland China) and *Walker* (Xingzhe, 2012) directed by Tsai Ming-liang (Taiwan), to examine their cinematic demonstrations of "inutility" as aesthetic and political potential resistance to the representationalism of capitalism. Focusing on cinematic affect, Liao engages with the potential of uselessness to resist a cycle of investment and reward. Given the cultural contexts and stylistic parameters of "inutility" as it is deployed in these films, Liao situates *Useless* and *Walker* alongside other alternative films within the archive of the New Chinese Documentary Film Movement to explain how these documentaries contest but do not replace the progressive narratives proffered by state-sponsored propaganda and capitalist consumerist culture. Because they do not surpass these other narratives, and, thus, create a surplus value, asserts Liao, these films are able (temporarily) to resist the economic logic of capitalism. Focusing on themes of estrangement and sacrifice, Liao assesses the relation between character psychology and costuming in Jia's film and the relation between urban haste and the slowness of repetitive character movement in Tsai's. Through these cinematic articulations these films challenge the "representationalist" foundations and investment/reward logic of global capitalism through their articulation of "the inutile," which resists capitalist appropriation. However, Liao also notes that, although they resist and open new sites to contest the ubiquity of global capitalism, these films still risk being reappropriated by the logic of capitalism. Thus, in the final analysis, Liao invokes Dennis King Keenan's concept of cultural formations that "sacrifice sacrifice" to rethink how cinema might "inutilise inutility," multiplying the politics of uselessness and endless artistic (re)creation.

In Chapter 7, Darren Byler also considers affect and documentary in relation to embodiment in his exploration of disposability in Xu Xin's *Karamay* (2010). Byler explores the valence of the biopolitical concept of "disposability" in tension with the anthropological concept of "ritual" to argue that a refrain emerging from a close reading of embodiment in contemporary independent cinema in Reform-era China is an effect of rituals failing to provide the well-being they promised in the Maoist past. Through his analysis of the monumental documentary film *Karamay*, Byler describes how the long duration of a historical trauma is embodied by a group of Han and ethnic minority oil workers and how this embodiment relates to an affective atmosphere of failure for those on the margins of economic development and social justice in Chinese Central Asia. By looking closely at political speech and political embodiment in the film, Byler extrapolates how ritualized ways-of-being arising to the fore in Maoist China continue to form a deeply felt common affect for marginalized people enduring rapid changes in the built environment and economic structures of mainstream contemporary Chinese society. Specifically, he describes how the long duration of a historical trauma, injustice, and alienation flows through cinematic bodies in the film. He then considers how this ritual embodiment relates to an affective atmosphere of failure for marginalized groups in Chinese Central Asia. This affect functions between and remains in tension with "disposability" and "ritual" so that, despite their failure on an institutional level,

Karamay's intimate portrayals of ritualistic gestures and motions still hail viewers in an embodied *phronetic* struggle for existential stability.

Lee Yuen Beng, in Chapter 8, examines how a group of Malaysian Chinese filmmakers metaphorically contest state policies and commercial Malaysian mainstream cinema through the cinematic display of acts of eating. Over the years, Malaysian cinema has solely focused on the problems and lifestyles of Malays, catered to Malays, and featured Malay casts speaking in Malay. In effect, Malaysian cinema is a Malay-centric industry that has negated the cinematic presence of other Malaysian ethnic communities. This ethnocentricity is currently being contested by a select group of prominent Chinese Malaysian filmmakers. Their works have collectively led to the rise of the Malaysian Sinophone cinema through their shared desire for the articulation and recognition of "Chinese-ness" in Malaysian cinema. In reversing the cinematic negation of traditional Malaysian cinema, they employ transnational methods to narrate stories of Chinese Malaysians through previously unheard of Sinophone films that have similar narratives and themes, delve into related subject matter, and deploy Sinophone languages. This collective usage of "Chinese-ness" goes beyond its cinematic representation by directly contesting and renegotiating Malaysian national identity while criticising the political economy of Malaysia and its cinema. This contest is carried out through scenes of eating, which are in essence depictions of the need to satisfy a hunger—a desire and longing for equal identification and recognition as Malaysians, part of the nation of Malaysia, no longer marginalised by race and ethnicity-based policies.

In Chapter 9, Andrew Grossman turns toward an analysis of "Category III" Hong Kong horror films, genre, censorship, and subversion within a Marcusian framework. This chapter tests the limits of allegedly subversive genre filmmaking and questions whether commercial horror cinema can legitimately claim to be oppositional. While acknowledging Herbert Marcuse's argument that allegedly "subversive" texts are usually themselves the products of an oppressive culture industry, the chapter nevertheless finds fugitive glimmers of subversion in select artifacts of Hong Kong horror cinema that fail to be recuperated within the limits of genre. Cultural studies often focuses on the marginal, less surveilled corners of underground horror and exploitation cinema as avenues of shadowy resistance and subversion, as sites where filmmakers and audiences can engage libidinal desires rendered taboo in mainstream cinema. If we believe the arguments of Marcuse, however, such subversions are themselves products of an overlording cultural repression that grants fleeting, licensed moments of apparent transgression as part of the culture industry's totalized plan of domination. These alleged subversions are thus inauthentic, mere symptoms of the industrialized repression that, in Marcuse's view, we habitually mistake for freedom. That most horror and exploitation cinema still reiterates heteronormative tropes and expectations apparently confirms Marcuse's view that what passes for "subversion" is hardly subversive. Without wholesale revolution and un-repression, it may initially seem we cannot escape from Marcuse's cultural trap. Nevertheless, there are rare moments when stealthy cinematic subversions are authentic (and not mere symptoms of the repressive bargain) because they violate not only mainstream norms but the heteronormative expectations of exploitation cinema. Focusing on the phenomenon of Hong Kong's "Category III" films, this chapter explores cinematic texts that sit on the borderline between symptomatic repression and true subversion and texts that cross the threshold, offering politico-sexual subversions unexpected by even exploitation cinema's own audiences. While these films are unlikely to exert noticeable effects on culture at large, they do provide important clues as to how authentic subversion might surreptitiously operate within texts otherwise beholden to the rules of genre.

Katrien Jacobs, in the final chapter, brings all our studies under closer consideration and brings the boundaries of our very subject—cinematic failure—into question. Chapter 10 explores feminine pornography as a type of eroticism and "art of failure" by looking at Asian and American women's experiences with digital networks and sexually explicit media in order to demonstrate how pornography and erotic cinema are sensed and consumed within a post-cinematic framework of overindulgence and cultural affect. This chapter outlines a post-cinematic and trans-Asian framework of pornography consumption that allows women to browse and get turned on by miscellaneous and contradictory models of embodiment and cultural affect. It compares the reactions of Hong Kong Chinese, Japanese, and American women to scenes from a Japanese hard-core movie, *A Young Wife Violated before Her Husband's Eyes*, and a Hong Kong "Category III" movie, *3D Sex and Zen: Extreme Ecstasy*. Through this analysis, the chapter shows that women regard pornography with a "drifting gaze." They can be ambivalent toward and easily aroused by fluctuating genres because they treat pornography as an "art of failure" and ignore the rules of arousal in hard-core graphic post-cinema.

* * *

In the end, it is possible to suggest that Thai filmmaker Apichatpong Weerasethakul's 2004 *Tropical Malady* (*Sud pralad*) is a film that essays much of what concerns those of us whose work appears in this volume, those of us who work at the intersection of film studies and body studies in a transnational Chinese context. In *Tropical Malady* a young soldier, Keng (Banlop Lomnoi), falls in love with a young man from the country, Tong (Sadka Kaewbuadee). Through the first half of the movie, we watch their relationship develop around simple, well-lighted meetings in town, sharing and trading everyday objects, the singing of Thai pop songs, and small gestures and embraces that navigate among their passions, their uncertainties, and their bodies. At one point the two enjoy a quiet moment sitting on a *sala* in the forest. Keng asks Tong if he can rest his head in his lap. Tong says, "No. . . ." Keng is disappointed and pulls away. Tong continues by explaining he was about to say "No problem," but he hesitated. As Keng rests his head in Tong's lap, he tells Tong that when he gave him that Clash cassette, he forgot also to give Tong his heart, so now he is sending it to him. He reaches to touch Tong's shoulder. After a moment, Tong says he can feel the gift of Keng's heart coming to him. Soon, a woman appears and asks if they want to buy flowers. In another scene, we see the two young men watching a movie in a theater. We never see the images they are watching but hear the soundtrack while the camera cuts between different views of the theater audience. As Keng and Tong begin to relax and enjoy the film, Keng puts his hand on Tong's knee and slides it up his inner thigh. Tong responds by trapping Keng's hand between his legs. They laugh quietly as they tussle erotically until Tong puts his arm around Keng's shoulders and Keng grabs his hand. The scene cuts to Tong in the men's room, flirting with another man. In a third scene, Keng watches Tong urinate on the side of the road, and when Tong walks back to him, Keng grabs his hand and begins smelling and kissing it. Tong protests that he has not washed his hands, but Keng continues. Then, Tong takes Keng's hand and begins kissing and licking it. He stops, lets Keng's hand fall, smiles, turns, and walks away. The film never simplifies their relationship but continually depicts the difficult negotiation of their corporeality and their desire.

Then, at the half-way point, *Tropical Malady* "drifts" through a minute-long fade to black from this difficult courtship tale to a more fantastic look at the entangled relation between the separate but mutually constituting material and immaterial aspects of embodiment. In "Spirit's Path," the title of the second half of the film, the soldier (Banlop Lomnoi) pursues the shaman/tiger (Sadka Kaewbuadee) who has been killing cows and terrorizing local farmers. (We are never quite certain if these two characters are or are not Keng and Tong again, but the two stories reflexively comment upon one another.)

As the soldier struggles to catch the tiger/spirit, he loses communication with headquarters, runs out of provisions, and becomes evermore frustrated and agitated. Eventually, as the film further explores the relation between bodies, thoughts, emotions, and drives in more poetic ways, a baboon approaches the soldier and speaks to him. The baboon cautions the solider that the soldier does not understand the complexity of his relation with the shaman/tiger, who considers the soldier his "companion" and his "prey." When the soldier and the tiger/spirit meet, they attack one another. The shaman/tiger wins, steals the soldier's rucksack, and rolls him down the side of a hill. At the bottom of the slope, the soldier grabs his rifle again and points it at "the monster" but does not fire. A short time later, the soldier mistakenly shoots a cow. As he watches, the ghost of the cow rises from the corpse and walks into the jungle. The solder gestures toward the specter and mutters, "Wait." Earlier the baboon had told the soldier there are two outcomes possible in this situation. If the soldier kills the tiger/shaman, the soldier will free him. If the soldier lets the shaman/tiger eat him, the soldier will join him. One night, as the solider becomes more frenzied, he begins to crawl through the jungle, thrashing at the underbrush and rolling in the dirt and leaves. He stops and shines his flashlight. On a tree branch above him, he sees the tiger/spirit. They stare at one another. The soldier brandishes his knife. The spirit/tiger speaks to him:

> And now…. I see myself here. My mother. My father. Fear. Sadness. It was all so real…so real that…they brought me to life. Once I've devoured your soul, we are neither animal nor human. Stop breathing. I miss you…soldier.

The camera pans the dark jungle and cuts to a drawn image of a tiger in a tree, his exceptionally long tongue reaching to a soldier on the ground below him. We hear the soldier's voice on the soundtrack: "Monster, I give you my spirit, my flesh, and my memories." The film cuts to a close-up of the soldier's face. Tears stream down his cheeks. The voiceover continues, "Every drop of my blood sings our song. A song of happiness." The film cuts to a shot of trees blown in the wind. "There…. Do you hear it?" asks the soldier's voice.

Yet, crucially, *Tropical Malady* is not *about* embodiment and the cinema. It essays the very contact zone where we touch the cinema and the cinema touches us. As

Jihoon Kim perceptively notes in "Between Auditorium and Gallery," encountering Apichatpong Weerasethakul's films means encountering an immersive space.[13] It means becoming engrossed in the films, in their affective duration, their dilation of time, their spatial dynamics that maintain our awareness of screen time. The films follow the rhythms of bodies—not the rhythms of narratives, shots, or cuts—so we do not so much watch or listen to them as experience their effects. I would add, Apichatpong's films, especially *Tropical Malady*, draw us into their mythology, entangle us in their mythmaking, by detaching memories from individual characters or situations and provoking us to relive them, even if we never experienced them in the first place. Ultimately, thinking through such encounters is what we talk about when we talk about the intersection between cinema studies and body studies, here in a transnational Chinese context, a concern with cinematic affects, maladies, monsters, and memories, with spirit and flesh, and the failure of practices to contain or exceed these encounters.

Notes

1. Sheldon Lu, "Notes on Four Major Paradigms in Chinese-Language Film Studies," *Journal of Chinese Cinemas* 6, no. 1 (2012): 15-25.
2. Ibid., 21.
3. Vivian Sobchack, *The Address of the Eye: A Phenomenology of Film Experience* (Princeton University Press, 1991) and *Carnal Thoughts: Embodiment and Moving Image Culture* (Berkeley: University of California Press, 2004).
4. Sobchack, *Carnal Thoughts*, 2.
5. Laura Marks, *The Skin of the Film: Intercultural Cinema, Embodiment, and the Senses* (Durham, NC: Duke University Press, 2000), and *Touch: Sensuous Theory and Multisensory Media* (Minneapolis, MN: University of Minnesota Press, 2002).
6. Chung-Ying Cheng, "On the Metaphysical Significance of *Ti* (Body-Embodiment) in Chinese Philosophy: Benti (Origin-Substance) and *Ti-Yong* (Substance and Function)," *Journal of Chinese Philosophy* 29, no. 2 (June 2002): 156.
7. Sobchack, *Carnal Thoughts*, 4.
8. Ibid.
9. Cheng, "On the Metaphysical Significance of *Ti* (Body-Embodiment) in Chinese Philosophy," 146.
10. Sobchack, *Carnal Thoughts*, 3.
11. Stella Gonzalez-Arnal, Gill Jagger, and Kathleen Lennon, introduction to Embodied Selves (New York: Palgrave Macmillan, 2012), 6.
12. Sobchack, Carnal Thoughts, 1.
13. Jihoon Kim, "Between Auditorium and Gallery: Perception in Apichatpong Weerasethakul's Films and Installations," in *Global Art Cinema: New Theories and Histories*, ed. Rosalind Galt and Karl Schoonover (Oxford: Oxford University Press, 2010), 125–41.

CHAPTER 1

The Ruined Bodies of Transnational Chinese Cinema
Brian Bergen-Aurand

Abstract

This chapter engages with questions of film ethics with regard to the ruined bodies of transnational Chinese cinema. As Olivia Khoo asserts in writing about filmmaker Wong Kar-wai's *In the Mood for Love* (2000), the body in recent Hong Kong art cinema "can no longer be regarded as a dependable marker of identity, whole or fully present, since Chineseness also appears today in increasingly fragmented forms tied to diasporic experience." The current situation, according to Khoo, where attempts to essentialize or reify this disappearing or phantasmagoric body prove to erase it further, calls for a reconceptualization "of an ethical diasporic Chinese viewing position out of the impossibility of rooted Chineseness." Rather than further trapping or fixing the image of the Chinese body (and especially the diasporic Chinese body) by making it permanently "visible" and putting it on display, Khoo argues we should rethink the ethics of neighbors and locality in these films and the spectral bodies they imagine. Following Khoo, I argue that the focus of transnational Chinese cinema on ruined bodies/bodies in ruins further provokes a reconceptualization of the relation between film ethics and film bodies with regard to the ambivalent signification of ruins. While ruins mark collapse, destruction, decay, and bankruptcy, they also mark experience, history, memory, culture, preservation, heritage, and inheritance. Thus, rethinking ruined bodies/bodies in ruins, foregrounds a relocating of the wounds and scars at the origins of diasporic bodies, displacing diasporic ethics away from origins in times and places toward acts of leaving or being forced to leave: an ethics of bodies relating in transit. This dissolution of origins is a significant part of the provocation of such transnational Chinese films as *Enter the Dragon* (1973), *Ju Dou* (1990), *Farewell My Concubine* (1993), *East Palace, West Palace* (1996), *Happy Times* (2000), *In the Mood for Love* (2000), *Be with Me* (2005), *Three Times* (2005), *Spider Lilies* (2007), *Lust, Caution* (2007), and *Uncle Boonmee Who Can Recall His Past Lives* (2010).

Keywords: Ruins, Embodiment, Transnational Chinese Cinema, Diaspora, Ethics

The Ruined Bodies of Transnational Chinese Cinema

Standing before him was China's history, the history of the last decades, or even of the last centuries or millennia. The endless arrogance and revolt, dissipation and vice; the rivers of blood and mountains of bones; the sumptuous yet desolate cities, palaces and tombs; the thousands upon thousands of horses and soldiers mirrored against the huge canopy of the heavens; the axe on the execution block, dripping with blood; the sundial with its shadow revolving around the glossy stone slab; the thread-bound hand-copied books piled in dusty secret rooms; the long, mournful sound of the night-watchman beating his wooden rattle . . . all these together formed these desolate ruins. However, history would not stop at this scene of ruin, no, it would not, it would proceed from here, and go on into the wide world.

—Bei Dao, "In the Ruins"

All These Together Formed These Desolate Ruins

In the closing moments of Wong Kar-wai's 2000 film, *In the Mood for Love*, Chow Mo-wan (Tony Leung Chiu Wai) appears at Angkor Wat in Cambodia, where he whispers his secret into a hole in one of the stone walls of the ruined temple. As Mo-wan leans toward the wall, his mouth murmuring inaudibly into the opening, the camera pans and tilts around him, and the film's haunting theme song fills the otherwise silent space surrounding his ruined body/body in ruins. The final act of a film built on lingering, lavish, saturated, decomposing images, and a saturated, phantasmal soundtrack, then has Mo-wan walk through one of the temple's hallways and exit, in shadow, both the frame and what remains of the narrative, through a doorway. The film ends in ruins and leaves us in ruins.

After Mo-wan leaves the screen space, the camera holds the frame-within-a-frame image of the temple at dusk seen through the doorway Mo-wan has just exited. In the subsequent set of images the camera travels aimlessly, unhurriedly across the now vacated premises: through its hallway interiors and around the ruined exteriors, briefly pausing at the now mud-and-grass covered hole where Mo-wan has whispered his secret. All we are left with are decomposition and ruins—the traces of Angkor Wat, the mud daub seal, the shadows of a fading sun, the secret whispered into the wall.

Yet, as Olivia Khoo argues in her essay, "Love in Ruins: Spectral Bodies in Wong Kar-wai's *In the Mood for Love*," the temples of Angkor Wat are not the only ruins in the film. According to Khoo,

> In *In the Mood for Love*, Wong presents an image on the screen that is (lovingly) detailed, dense, and saturated, and yet on the edge of ruin and decay. The specter of an almost ungraspable love is nevertheless made present at the level of film form and style in what I call a spectral aesthetics.[1]

In the Mood for Love is not so much composed as decomposed. Architecture, temples, sounds, living arrangements, costumes, skin, hair, eyes, and bodies in general are in ruins in the film. They are the sumptuous yet desolate elements of the film's decomposition before our very eyes, the ruins of the arrangements of the film already under threat of ruination, of collapse and dissipation, ever more vulnerable because they expose a ruination at the center of arrangement—a decomposition at the center of composition—a ruination that marks arrangement throughout the form and content of this film and throughout the form and content of a certain segment of transnational Chinese cinema in general.

History Would Not Stop at This Scene of Ruin but Go On into the Wide World

Our experience of ruins is always ambivalent and embodied. "Ruins," as Khoo writes, are remainders and reminders of things past; they provide evidence of the past, authenticate it. The fact that ruins are by their nature fragmentary invites viewers to imagine or reconstruct a lost whole from the enduring fragments; they also reflect the ethnic spectator's desire for a whole body, an authentic presence, to be mirrored on the screen—despite this impossibility.[2]

Ruins function in terms of memory, history, and imagination. When we encounter ruins, we imagine them as enduring fragments of a larger, more complete whole that has been lost over time—a complete whole we seek to recover. Thus, in a gesture of return to this imagined whole, our encounters with ruins provoke us to remember, to locate them within our sense of history, to reconstruct a past when this imagined, authorized whole existed unbroken. Except, this whole is a product of the very ruination that created the ruins we encounter; thus, the very whole we seek to recover is impossible not because we cannot return to it but because it never existed in the first place. Ruination in this sense is not aligned with "melancholic grandeur," "nostalgia." It is a fantasy projection—like a film—that reflects back our own desires for authenticity. Like mirrors—rather than windows onto the past—ruins reflect back our desires as if they were authentically projected from the other place and time of the ruins to measure our authenticity with regard to them. Ruins provoke us to ask the impossible question of whether or not we are authentic to our roots. (This affect is why some writers have written of the "uncanny" experience of ruins.) In provoking us to consider our present situation in relation to the past, ruins prompt us to fantasize about the future, about the outcomes of these past expectations as they transition through our present toward the future. Encountering ruins, then, we encounter reminders of the same historical forces, the same forces of ruination that have formed us and, as the narrator in Bei Dao's story (from which the epigraph of this chapter is drawn) explains, would not stop at this scene of ruin but "would proceed from here, and go on into the wide world."[3]

In English, the noun "ruin" recalls "collapse, fall, and dishonor." It comes from the late fourteenth century and originally referred to the "act of giving way and falling down," from the Old French *ruine*, from the Latin *ruina*, "a collapse." It is related to the Spanish *ruina*, Italain *rovina*, and to *ruere*, "to rush, fall violently, collapse," which is of unknown origin. "Ruin," meaning the "complete destruction of anything," is from the 1670s, and "ruins," meaning "remains of a decayed building or town," is from the middle fifteenth century. "Ruin" as a verb is from the 1580s, and with the financial sense of "bankrupt" is attested to from 1660 on. This latter, economic sense is also related to carnal associations with "ruin" from 1567 (if not before). To be ruined, of course, is somatic in regard to health and honor. An unhealthy person is said to be "ruined"; a woman who has lost her virginity in an improper manner (seduced and abandoned) is said to have been "ruined." Thus, ruination has never been just about finances, structures, and architecture. It has always also been about the economics of bodies.

Ruination is a collapsing of the past, present, and future, as ruins bring us from the present into relation with other times. Ruins remind us how these times fall into, fold into one another, and how the collapse of ruination prompts questions of the ethics of failure, where "failure" signifies in the sense Christopher Pavsek describes in *The Utopia of Film*.[4] Pavsek writes of failure as possibility, not because the future might be made to match a past success or ideal but because the future is exposed to what has not succeeded in the past, and ruins mark not the success of the past but its failure, and thus a potential possibility for the future. This conception of ruins also shares a certain affinity with the rejection of "nostalgic fantasies" analyzed by William Schaefer in "Shadow Photographs, Ruins, and Shanghai's Projected Past,"[5] where Schaefer explores displaced and displacing images, relics, and texts from Shanghai's dislocated past in relation to modernity not to discover what really happened but to describe the process by which the past dually is transmitted into and haunts the present.

While ruins are ancient, historic, preserved, memorial objects, they also function as objects of subjective appropriation and mastery, whether epistemological or aesthetic. As Mo-wan appropriates the ruins at Angkor Wat to bury his past secrets, he preserves them for the future. As he buries what has been, he makes a promise to what is to come, a future that will discover his ghostly trace and be brought into the ruination of his having been there but not being there now. His disappearing body, already fuzzy in the middle distance of the shot, disappears as he exits the frame. Mo-wan is in transit here, in Cambodia, in this film, transitioning through the collapse of place and time, collapsing before our very eyes.

Thus, ruins, to be a bit more schematic, are matters of form and content and in films such as *In the Mood for Love* and other transnational Chinese cinema affect all three gazes of the cinema through *a ruined gaze*, in effect to ruin the gaze to know what is really happening or what really happened. Since the structuralist turn of film theory in the 1970s, it has been standard practice to discuss cinema in terms of three gazes or looks: the gaze of the camera in recording the scene present before it; the gaze between/among the characters present in this scene, which is then projected on to the screen during the film's exhibition; and the gaze of the spectators/viewers present at that exhibition. Behind these cinematic gazes lies not a window on the world but an ideologically constructed life-world, which we are taught to believe in and desire. The three gazes of the cinema work to arrange and position us to their point of view. The ruined gaze, such as we get in *In the Mood for Love* and other transnational Chinese cinema, looks backward to highlight the arrangements of the three gazes of the cinema. It works to ruin their arrangement and positioning by dislocating and displacing them. At the same time, through its failure in the present to render a clear sound and image of what is happening and to present success stories or successful bodies, it ruins the present. Thus, it remains transient as it looks forward toward an ethics of failure located on ruined bodies/bodies in ruins brought to light through the cinema.

Imag(in)ing the Ruins of Transnational Chinese Cinema

It is a question of how we imagine ruins. If ruins are always located in time and place, if their very ruination depends upon their location, then how we imagine them—a variety of dislocation or relocation of ruins—becomes especially important with regard to an ethics of failure in transit. For, in imagining ruins through the ruined gaze we imagine images of ruins through images in ruins. In the transnational Chinese cinema under consideration here, it is not a question of representations of ruins alone. These films do deploy sounds and images of ruins. Yet, they also do more than that. They enact a certain ruination on the bodies within these images but also upon the images themselves. The ruined gaze looks upon ruins, rehearsals, deletions, displacements, decompositions, marred bodies, wounds, and scars. It eschews the completed, whole, and sharply focused. It displays journeys or experiences—emotional, psychological, political, ethical, philosophical, and aesthetic ones—with regard to coincidence and intersection rather than proper actions and plans brought to fruition. Ruins haunt these films as ghost images, specters, disappearing images, displaced images, phantom images, images of phantasms and phantasy. They function as indirect images—hallucinations (seeing what is not there) and reverse hallucinations (not seeing what is there), and as underexposed and overripe images.

Of course, not every instance of transnational Chinese cinema deploys every one of these techniques of ruination; nor does any one film use them all. Yet, they all share in these tropes and compositional techniques in one way or another. Or, we might claim that these films have a tendency toward these elements—toward ruining the gaze—even when they do not fully decompose their sounds and images. Their narratives focus on missed encounters and characters who arrive too early or too late, who are always just slightly out of time or out of place. These characters are haunted by loss and failure, especially of pledges and promises they cannot keep, and the relationships lost to these failures. Much of the films exist outside the film frame with characters inhabiting the periphery of the image, focusing their looks somewhere off screen and their gestures and verbal remarks to something or someone unseen, signified only by the gesture or remark directed toward them. When the characters are on screen, very often their costuming is more stylized than realist, often with an exaggerated shimmering effect created especially by formal wear and Cheongsam (*qipao*) dresses or an excessively drab sartorial atmosphere produced by military or institutional attire. In many of these films, costuming almost overwhelms the bodies it contains, rarely fitting properly. Rather, attire is tight and restricting, binding in place the shape of the body, or it is loose and flowing, overwhelming and concealing. And then, in a flash of violence is torn away, revealing the flesh and (very often) the blood of a wound.

The films are shot in dense, oversaturated color, but the shadows are blurred, flattening or blurring the planes of the image. The palette is a combination of earth tones and bold primary color, often set in sharp contrast. The images are rarely garish. Light is rarely direct—almost always filtered through screens or atmospherics. Even when it appears to come from a diegetic source, it creates a glow and halo effects around characters and objects. There is, in many of these films, what Khoo calls an almost suffocating beauty of shimmering and swirling shapes and colors.[6] They are overripe visually and aurally. The sound is also saturating—whether ambient or orchestrated. The music seems to be both diegetic and non-diegetic simultaneously, and the films deploy their music for more than emotional or thematic enhancement. Sound overall plays its own part in these films, sometimes overwhelming the image rather than serving it. When voice-over is used, it is melancholic or overdetermined. As with the images, the sounds of these films decompose more than they compose. They "disquiet" the films in the sense that they interrupt the peace or tranquility of scenes, the passivity of scenes. In a similar way, the films are filled with anachronistic techniques that decompose the divide between fantasy and reality, collapse the times of earlier periods and contemporary settings without returning the investment in anachronism.[7] The films bank on this display of decaying time frames through slow-motion shots, stop-action photography, unmotivated flashbacks and flash forwards, narrative lines from different eras woven into one another, and characters and actors appearing and disappearing throughout different time frames, sometimes as the same persona, sometimes as a different persona.

Finally, it is a question of the location of ruination, the timing of ruined bodies/bodies in ruins, of failure, and transient ethics. What tends to occur across these films is a certain regard for the importance of ruined bodies/bodies in ruins coming at crucial moments in the narratives. Intellectual, emotional, psychological, sexual, financial, political, ethical, and aesthetic failure tends to be traced on the bodies of these characters at crucial moments in the films. Their "fall" in immaterial ways is made material through the marks, bruises, wounds, and blood on their bodies. Their experiences (traumatic and otherwise) do not simply haunt them or remain latent in their memories and histories. They are written on the body—often leaving scars and tattoos—and most prevalently when their bodily ruination is a manifestation of change in situation.

Ruined Bodies / Embodied Ruins

"Everybody loves Bruce Lee's body, or so it seems. But they may not all love it for the same reasons," writes Chris Berry in his essay on Lee's corporeality, the intersection of Chinese and American masculinities, and the genre of martial arts films and transnational cinematic frameworks.[8] As Berry and others have noted, Bruce Lee altered our perception of the Chinese body through the films he made between 1971 and 1973. Prior to Lee, martial arts films did not focus primarily or exclusively on male protagonists, often emphasized a refined, gentlemanly, or scholarly masculinity when they did deploy masculinity, and displayed characters fully clothed—from head to toe—in loose fitting, flowing attire that shrouded rather than emphasized the body. Lee's self-display challenged these expectations, and by putting the exposed martial male body on display, opened our field of vision to different images of masculinity. By visualizing the male martial protagonist in ever-more revealing states of undress and muscularity Lee exposed the American masculine "hard body" to the cinematic gaze while his characters also retained a certain refined masculine disdain of power; Berry labels this hybridization "Lee's modern and transnational neo-*wu* masculinity."[9] In each of his films, Lee strips off his shirt—and sometimes other articles of apparel. In each subsequent film, this disrobing occurs earlier in the narrative than it had in previous films, affecting our senses of the relation between embodiment and exposure as costuming.

While Berry stresses how Lee's redeployment of neo-*wu* embodiment invites a queer gaze and queer viewing pleasure and in the process produces a homophobia the films must jettison through their disdain for soft male bodies, Lee's remasculinization also puts on display the exposed male body—often after it is wounded and bleeding—the male body in pain, the male body in ruins. In his final film before his death in 1973, *Enter the Dragon*, Lee appears almost nude from the start of the film. He is an exceptional practitioner of martial arts—a teacher, not a student—and already an insider in the world of transnational espionage. Lee plays a secret agent working for MI-5 who infiltrates a martial arts tournament held on the island fortress of the evil Shaolin master Han. Han is marked by a prosthetic left hand, which he replaces with claws and knives at different points throughout the film. As with most films in the genre, embodiment functions in regard to ethics, puts morality on display, and Han's dis/ability marks his dis/honorability, as he fights deceitfully, using his prosthetic weapon to his advantage in "unarmed" encounters.

Han functions precisely as Sharon Snyder and David Mitchell describe in their study of body genres and disability sensations in *Cultural Locations of Disability*. According to Snyder and Mitchell, "body films attempt to situate the filmed body in the throes of excessive emotion as an object of mediation for an anticipated viewer's own experience of embodied peril."[10] These spectacular, dis/abled, filmed bodies are put on display to caution filmgoers of their own boundaries, emphasizing "a social investment in certain bodies' presumed proximity to abjectness."[11] This spectacle associates dys-functionality, pathology, and dis-integrity with ethical decay and ruin. The display of Han's deceptive prosthesis displays Han's deceptive spirit. As Snyder and Mitchell explain further, the offensive body is the body "gone awry":

> Bodies must remain within certain boundaries, and their "leakage" beyond such parameters violates social expectations of propriety (the appropriate self-mastery of one's bodily functions, fluids, and abilities).[12]

Furthermore, near the end of *Enter the Dragon*, Han's body "gone awry" is deployed against Lee's integral body in the final fight scene. This sequence of ruination at the end of the film sees Han adopt spiked and then bladed prostheses he uses to flay Lee. The initial armed attack is surprising and shows a trickle of blood on Lee's face. Subsequent attacks weaken Lee's bodily integrity and let more blood. Han (also bleeding from the nose and mouth) manages a cut across Lee's midsection. Here, Lee touches the wound and tastes the blood on his own fingertips. The battle continues as Lee follows Han into the bath/closet of mirrors, where Han cuts into Lee's bare and bleeding torso several more times. The mirrors disfigure, disorient, and disseminate the men's bodies, and Lee's shattering them fails to reduce this spectral effect. Lee defeats Han with a kick that pushes Han onto his own spear projecting through the revolving door. The scene ends with Han's penetrated body spinning on the door as Lee leaves the room.

In a way, this chapter thinks transnational Chinese cinema through the wounds and blood of Bruce Lee's body, an already transnational, diasporic body in transit (as many critics have analyzed). Lee was born in Chinatown, San Francisco, to parents (of Chinese and German extraction) from Hong Kong. When he was young, Lee returned to Asia with them and began work as a child actor. At the age of 18, he attended the University of Washington. After leaving school, Lee taught martial arts and began a television career in the United States, most famously co-starring as Kato in *The Green Hornet* (1966–1967) and *not* starring in *Kung Fu* (1972–1975).[13] In 1971, Lee returned to Asia to make Hong Kong and Hollywood produced films that became international blockbusters. All the while, he held dual citizenship from the United States and Hong Kong.

One could certainly compare and contrast Lee's career and bodily display to those of other transnational Chinese stars, such as Jackie Chan, Jet Li, Stephen Chow, Chow Yun-fat, and Michelle Yeoh. One could certainly consider the role of ruined corporeality and desire across the vast body of transnational martial arts and related films, especially *Drunken Master* (1978), *Police Story* (1985), *Once Upon a Time in China* (1991), *Crouching Tiger, Hidden Dragon* (2000), *Hero* (2002), *House of Flying Daggers* (2004), and *The Grandmaster* (2013). Such a basic comparison would continue to open new discussions in genre and body studies. However, comparing and contrasting Lee's deployed embodiment differently, in relation to the displays of other bodies in other film genres and across other cinematic movements alters the very function of transnational Chinese bodies on screen. Deploying Bruce Lee's ruined body / body in ruins in relation to these other bodies complicates the address of the ruined embodiment of these other films and alters the parameters for discussing all transnational Chinese cinema.

At the end of the 1980s and beginning of the 1990s, three films starring Gong Li and directed by Zhang Yimou traveled around the globe and received a vast quantity of international recognition. Many critics have written about the extravagant exotic display, focus on desire and repression, and institutional critique of the films. Many others have also questioned the films and filmmakers over the target and intention of their display. In one chapter of *Primitive Passions*, Rey Chow addresses the "double gaze" and excessive surface imagery of Zhang and Li's first three films— *Red Sorghum* (1988), *Ju Dou* (directed with Yang Fengliang, 1990), and *Raise the*

Red Lantern (1991), arguing that "what Zhang is producing is ... an exhibitionist self-display that contains, in its very excessive modes, a critique of the voyeurism of orientalism itself."[14] This ruinous excess is the force of surfaces in these films, according to Chow.

Following and developing Chow, one can see the (ruined) bodies deployed throughout these films (and *The Story of Qiu Ju* [1992]) return a ruined exhibitionist gaze to spectators precisely through their excessive ruination of the ruined bodies/bodies in ruins on screen. In *Ju Dou*, against the backdrop of luxuriant vibrant colors in the dye factory, the impotent Yang Jinshan beats two wives to death for failing to produce heirs. Yang Jinshan buys Judou as his third wife and sexually abuses her each night; here it is the flow of her blood that is important, of course. Eventually, Judou, in the shadows of her bathing room, turns to expose her bruised and beaten body to Yang Tianqing—the nephew who has fallen in love with her. Judou and Tianqing produce a child but pretend it is Jinshan's heir. Jinshan has a stroke and begins using a wooden barrel cut in half and set on rollers for mobility (again the villain displays a spectacularly dis/abled body). Discovering the biological lineage of Tianbai, Jinshan tries to drown the child. Tianbai survives, but Jinshan later drowns when he falls into a dye vat. Judou and Tianqing prostrate and ruin their bodies as part of the long and demonstrative funeral procession. Ten years later, Tianbai discovers Tianqing is his biological father and drowns Tianqing. In the final act of ruination, Judou's ruined body stands in the middle of the ruined dye factory as she burns it to the ground. We lose sight of her as the flames surround and engulf her dirty face and weakened body.

In a similar way, *Raise the Red Lantern* also addresses ruining the gaze by putting the excessively ruined body on display in so many of its shots—alternating between long shots and close-ups of characters so often seen staring—framed by doorways, hallways, and window openings. In this film about multiple wives held in a prison-like house, the ending is of special concern as the beauty of the bodies deployed throughout the film and the excessive display of costuming, color, and lighting that has dominated the mise-en-scène are disrupted by the closing scenes. The top of the house compound is covered with snow. All is white or dark gray. As Songlian approaches the ancient room—which the men must force open—where Meishan has been hanged, a rare medium shot shows her moving across the rooftop. The film cuts to a point-of-view shot of her looking at the locked door and then cuts to a long shot as we see her peeking into the chamber. Her screams begin to disquiet the previous silence of the audio space. Later, the red lanterns have been relit in

Meishan's room, and the men fear a demon has come to haunt them, but the room is empty of any body. The red glow from the lanterns bathes every surface in the space. Suddenly, the film cuts to Songlian lighting all the lanterns in her room. She brings Meishan into the space through the recording of her voice she plays on the record player. The red glow bathes all of Songlian's room and seeps out through the doors and windows. The next summer, the Fifth Mistress is brought to the Chen house. The film closes with Songlian, dressed in her original school uniform. Braids unkempt and mouth agape, she stares blankly at the ground or into a middle distance as she paces her room and wanders her courtyard. The still-lit red lanterns hang about her. All else is gray or black.

In the story by Su Tong, on which the screen play is based, the narrator recounts that everyone in the household wondered: "But why would the young, perfectly healthy, and refined Fourth Mistress Lotus suddenly lose her mind?" and a few lines later, "Bamboo [the Fifth Mistress] saw that she was very clean, pretty, and refined, not at all like a madwoman."[15] The film, however, displays the assumed somatic analogs of being cognitively and emotionally dis/abled by the surrounding social structures. The image of Songlian's body is ruined and moves about the ruins of her space to reflect her cognitive and emotional ruination.

Unlike the previous films by Zhang and Li, *The Story of Qiu Ju* is set in contemporary China and addresses ruination that has already occurred before the film begins. Qiuju and her husband Qinglai live in a small farming enclave. She is pregnant, and her husband is injured and unable to work because he fought with his boss, who subsequently kicked him in the groin and severely injured him. With its realist style and subdued display the film is not as excessive or extravagant as the other films discussed in this chapter. Yet, its focus remains on the ruined body and the institutions that contextualize it. Qiuju struggles with the boss who injured her husband and with the system of, albeit sympathetic, officials who might address her situation. Nobody in the film is particularly mobile, and the film emphasizes the distances Qiuju must travel to have her case heard. The climate and geography offer passive resistance to her mobility as she remains in transit through almost the entire movie, juxtaposed with her husband who barely moves.

This image of a more mundanely restrained yet still transient female body intersects as well with Zhang's 2000 film *Happy Times*, where Wu Ying's blindness seems more en/abling than dis/abling of community building among the ruins of the condemned factory and laid-off factory workers. The film's two endings—one for domestic release and one for international release—juxtapose the ruined body and the body in ruins. While in one ending, Wu Ying and Zhao, who would have been her step-father had his plans to marry her step-mother not failed, sit together in the ruined factory reading a letter supposed to be from Wu Ying's biological father. The industrial ruins almost overwhelm the human figures on screen. In the other ending, Wu Ying has left in search of her biological father, and Zhao has been hit by a truck. The film shows him lying bandaged in a hospital bed and then cuts to his friends who have gathered at his apartment. There, they find a cassette tape from Wu Ying, who expresses her final thoughts and emotions in voice over as they listen to the recording. (Wu Ying is returned to the filmic space through her recorded voice, recalling Songlian's playing of Meishan's recorded voice on the record player.) One friend, Little Fu, in a remarkable image of prosthetic mediation and somatic distancing, reads the letter, which was actually written by Zhao in what he imagines to be Wu Ying's father's voice, to the sounds of Wu Ying's voice. In this version, prosthesis and ventriloquism replace the bodies of all three characters.

From the start, the different steps in the ruination of Douzi/Cheng Dieye's (and to a lesser extent, Shitou/Duan Xiaolou's) body mark the plot points of *Farewell My Concubine* (Chen Kaige, 1993), a film saturated with the sounds and images of somatic technique and the deployment of trained bodies. The film displays the story of these two Chinese opera singers and Master Yuan Shiqing (a patron who pursues Douzi) and Juxian (a sex worker who marries Shitou) against the backdrop of the ruination of twentieth-century Chinese history from the 1920s to the 1970s. As the narrative progresses linearly, after an initial flashback from the 1970s to start the story in the 1920s, we experience the ruins of these boys' bodies against the ruins of China following Japanese occupation, liberation, the consolidation of control by the Communists in 1949, and the Cultural Revolution, engaging in what Song Hwee Lim calls, "the double taboo of politics and homosexuality in China."[16]

When Douzi arrives at the opera school as a young child, he is initially rejected because he was born with a "superfluous" eleventh finger. His mother (a "ruined woman") immediately takes the boy outside and chops off the finger. She returns to sign the school contract in the boy's blood. In the eyes of the school master, Douzi's body was ruined from birth; in the eyes of onlookers, his body is ruined from the moment his mother violates its integrity in order to normalize and institutionalize it. For years, Douzi, Shitou, and the other boys are physically strained by their training and beaten by their teachers. Eventually, Douzi's body is made so docile by the systematic discipline and punishment that he turns to slapping his own face bloody when he feels responsible for underperforming.

After Douzi commits to becoming a Beijing Opera star alongside Shitou, he begins training to play *Dan* (leading female) roles across from Shitou's *Jing* (painted-face male) leads. Douzi will learn to play the concubine beside Shitou's Chu King—who is defeated by the Han king when his entire army deserts him. Douzi's first lesson is learning to recite the monologue "Dreaming of the World Outside the Nunnery," in which he must sing "I am by nature a girl, not a boy," but he cannot complete the oral gender inversion and instead repeatedly sings, "I am by nature a boy...." The school is under threat of closure, and Douzi has the chance to impress a wealthy patron, but still cannot sing against his corporeality. Here, Douzi's queer attachment to Shitou is made material as Shitou (angrily but with sympathetic tears in his eyes) penetrates and scrapes Douzi's mouth with a wood and brass opium pipe. As with the first moment at the school, Douzi's blood again corporealizes the narrative shift, as he stands and sings the song perfectly as blood flows from his lips. His bodily ruin has kept the school from financial ruin. Furthermore, after the later performance for the wealthy patron Douzi is summoned to the bedchamber of the Eunuch Zhang: "The concubine has to die one way or another," says the house manager to the teacher accompanying the boys. Alone with Douzi, Eunuch Zhang tells the boy to pee into a vase and then intimately assaults him. When Douzi leaves the chamber some time later, he is physically shaken, refuses to speak, and stares vacantly into the distance. His further bodily ruin has again protected the opera troupe from financial ruin.

The troupe adopts and trains a baby Douzi has discovered, Xiao Si, who also becomes a *Dan* singer in time and eventually displaces Douzi as the concubine beside Shitou. Shitou marries Xiaolou after several complications. Douzi cohabitates with Master Yuan for a time. These intimate events transpire against the background of the various transitions of Chinese governance mid-century. As the Cultural Revolution begins, the opera company is brought in for questioning and self-incrimination. Under duress, Shitou accuses Douzi of singing for the Japanese and having an affair with Master Yuan. Douzi exposes Juxian as a sex worker. Shitou denounces Juxian, who hangs herself upon hearing of Shitou's condemning her. Douzi becomes addicted to opium. Later, even Xiao Si is apprehended after opera is outlawed as reactionary and he is caught practicing his *Dan* role. The film ends in 1977, with a final rehearsal of the opera *Farewell My Concubine* with Douzi and Shitou. Douzi repeats his childhood error—"I am not a girl..."—and Shitou corrects him. In this empty, dusty, and underlit ruined theater comes the final ruination of Douzi's body as he commits suicide by sword, replicating the role he has been practicing and performing all his life, highlighting the performative aspects of embodiment (whether essentialist or essentializing or not) and the failure of the ruination of Douzi's body throughout to guarantee any proper action or plans come to fruition. All that remains is Shitou holding Douzi's ruined body in the ruined theater.

From the opening of the film encounters between and among characters are incidental, nearly missed, and out of time and place. Douzi and Shitou are surprised to meet in the ruined theater. Then, once the flashback begins, it is Douzi's amputation that brings them together, the accidental discovery of the baby Xiao Si leads to problems later in the films, and Shitou announces his engagement to Xiaolou only when the men attack her, claiming she is a ruined woman in the Blossom House. In this way, as much as the film narrative is composed through causal narration, it also decomposes any absolute reliance upon causality. The mise-en-scène is filled with shimmering costumes (emphasized against the drab military and institutional uniforms throughout the film world) and transient sets; while the opera actors, especially Douzi, often appear in sumptuous, if archaic, attire, the buildings and cityscapes are quite mundane. The old school appears dusty and about to collapse in places. The opera houses are smoke-filled, and dust rises in the streets in many outdoor scenes. One has the sense, when watching *Farewell My Concubine*, of seeing it through gauze or thick glass at times—an aspect made apparent by way of several shots through linen, paper windows, fish tanks, and screens. Finally, since it is a film about an opera that deploys the operatic throughout, the film emphasizes sound—especially the opera orchestra—that often saturates the visuals, overwhelms and disquiets the filmic image, combining with the other elements to displace the sense that this film portrays "what really happened." In his discussion of the film, Song concludes that,

> the use of [Douzi's] femininity as sacrificial death, therefore, can be seen as a nostalgia for a lost era, when monumental structures of feeling were highly valued and the role of the artist-intellectual vis-à-vis the state, albeit feminized and submissive, at least had a paradoxical potential to become heroic and super(hu)man through suicide.[17]

Yet, to consider Douzi's suicide a ruination, among many other ruins and ruined bodies, rather than a sacrifice against an arranged background such as the state, bends the ending of the film away from nostalgia toward an ethics of ambivalent transience. Seen against the backdrop of all the ruined institutions, architecture, and hierarchies displayed throughout the film, Douzi's self-ruination exposes the failure of the past and the present in the light of future possibilities. The question veers from one of historical authenticity to futural possibility: "What might happen?"

Addressed in this way, *Farewell My Concubine* intersects as much with other ruination films, such as *Shower* (1999) and *Happy Times* (2000), as it does with crucial transnational queer cinema, such as *East Palace, West Palace* (1996), *Fish and Elephant* (2001), the films of Tsai Ming-liang, and *Spider Lilies* (2007)—this last overtly connecting ruined bodies/bodies in ruins through the repeated image of a dead father's arm protruding from the rubble of a collapsed building and late scenes involving a young man taken to hospital after his arm is severed in a street fight. *Shower*'s display directly addresses the link between ruined bodies/bodies in ruins through its juxtaposition of carnal and architectural ruination, although, as Jami Proctor-Xu describes it,

> *Shower* marks a moment in which ruins become a legitimizing site for narratives that naturalize the destruction modernization brings even as it occurs. The film is a product of transnational capital that apologizes for the destruction resulting from the modernization via capitalist development and the erasure of traditional architectural structures and lifestyles from China's urban landscape. Old Liu's weakened body and the dilapidated bathhouse signify a cityscape and a way of life that the film in the end dismisses as already part of China's past, while the female body evokes merely nostalgia for what is lost in the process.[18]

Seen through the lens of nostalgia for the lost body and lost environs, the film appears to support an ideology of dismissing and forgetting the past. However, the film's paralleling of bodies and buildings might also recall the impossibility of a rooted Chineseness, not in support of transnational capitalism but of the ambivalence of transnational transience where diasporic and failed foundations connect to an ethics of bodies relating in environs to come, signaled in part through the relationship between Da Ming and Er Ming, his cognitively dis/abled younger brother whom he brings to live with him at the end of the film. Again, we can ask,

"What might happen? What might past and present failures make possible in the future?"

East Palace, West Palace (1996) was the first feature film from mainland China to directly address contemporary gay life in Beijing and is another transnational Chinese film addressing the embodiment of drama, rehearsal, performance, and performativity. The title refers to the two public toilets on either side of the Forbidden Palace on the north edge of Tiananmen Square, and it may be one of the few films where architecture remains a more stable background against which bodies are ruined—until the final moments of the film. Most of the action is set in the park police station, where A Lan is interrogated by Shi Xiaohua, and A Lan proclaims his love for Shi Xiaohua repeatedly. The mise-en-scène is tight and secluding; the park becomes an overripe jungle as the night progresses. Throughout the film, bodies are twisted and contorted as Xiaohua puts A Lan in stress positions during questioning. The two men struggle with each other at times. A Lan is bound and released. He is told to disrobe and put on a skirt and jacket. When he does, he also applies red lipstick to his lips and cheekbones. At first, the lipstick is smeared on his left cheek. A moment later, though, his makeup and costume are perfect. At the end of the film, the men rush from the clean, well-lit police station to a damp and decaying alley. Xiaohua says he will cure A Lan. A Lan protests he is not sick, only in love. Against a wall covered with peeling paint and torn posters, the men begin to kiss and embrace roughly. A Lan loses a shoe. Inside a dilapidated building, Xiaohua tears A Lan's clothes from his body and begins to douse him with a hose. A Lan falls to his knees and embraces Xiaohua around the waist. The men grow gentler, part, and A Lan watches as Xiaohua slowly walks away. In this closing sequence, the film displays the transience of ruined bodies/bodies in ruins while displacing the ethics of this encounter by decomposing an image of the future brought to fruition. The past and present lie in ruins, opening toward a future without roots, without guarantees.

Lust, Caution (2007) is notorious for its deployment of bodies—the deployment of graphically depicted, fully exposed sex scenes on screen and the deployment of the bodies of Tony Leung Chiu Wai and Wei Tang in those scenes. Ang Lee's adaptation of Eileen Chang's story is shot in an oversaturated, overripe color recalling to some degree Wong Kar-wai's *In the Mood for Love* and at least the visual style of Hou Hsiao-Hsien's *Three Times* (2005). It is a tense espionage drama about dramatic performance and gender performativity, set during the Sino-Japanese War. Wang Chia-Chih is a spy attempting to seduce Mr. Yee, a collaborator with the occupying army. The film emphasizes the psychology of espionage, the mood and affect of being a spy or being spied upon, and the fear of being discovered. It is a film very much concerned with desire and corporeality under "dispersed" surveillance as Foucault describes in *Discipline and Punish*.[19] Throughout, the film hints at the erotics of espionage and the embodiment at stake in impersonation—especially emphasizing the youth of the spies, a company of university actors who have banded together. In two scenes, however, the boundary between desire and corporeality collapses, and this collapse opens to the relation between ruined bodies/bodies in ruins and transient ethics.

Almost halfway through the film, the truth about the cell is discovered by an acquaintance, Tsao, whom they had been deceiving, and Tsao threatens to blackmail and expose them. K'uang Yu-min (the leader of the cell who is in love with Chia-Chih) challenges Tsao with a knife, and Tsao pulls a gun. The members of the cell overpower Tsao and begin beating and stabbing him. The murder is portrayed as an erotic and complex affair: the attack begins after the friend alludes to sexual payment from Chia-Chih; as three of the men hold Tsao on the couch, Yu-min spreads his legs and then straddles his waist to stab him in the stomach; the knife slips, and Yu-min slices his own palm; when he does stab Tsao (in a slow and deliberate penetration), it is not lethal. Tsao pulls the knife from his own stomach and attacks Yu-min. The men overpower Tsao again, and another of them stabs him three to six more times. Tsao and Yu-min's blood spreads throughout the room. Tsao does not die but rises from a pool of blood and walks to the door. A different member of the cell again stabs him twice in the back. Tsao turns and walks out the door toward the stairs. The same member stabs Tsao again in the back, Tsao—his white shirt soaked with red blood—falls face-first down the stairs. As the cell members look down from the top of the stairs one woman cries, "He's not dead yet!" Yu-min descends the stairs to straddle Tsao again, this time from behind, grabs his head and twists his neck. Tsao stares up at Yu-min as he dies. Yu-min looks up at Chia-Chih. Despite their previous doubts about the mission, they must now continue.

It has taken well over two minutes of screen time to kill Tsao, and this brutal scene makes visual—corporealizes—the psychology of the characters. Both "lust" and "caution" are deployed here through the ruination of Tsao's body, especially as it is sparked by Tsao's implication that he will extort partial payment for his silence through having sex with Chia-Chih. Indeed, it may be possible to suggest that this scene is precisely the ruination of the border between caution and lust, which have been established as oppositional desires through the first part of the film. Here, it is uncertain whether lust or caution triggers the assassination of Tsao or if they are separate desires at all. Perhaps, the scene suggests, lust and caution are intimately connected, the desire to survive linked to the desire to perish.

This dissolution of the border between lust and caution returns in a later conversation when Old Wu, to whom Yu-min and Chia-Chih report, meets with them to discuss Chia-Chih's progress with seducing Mr. Yee. In this highly charged scene, Chia-Chih reassures Old Wu of her loyalties to her party, her leader, and her country. Old Wu takes Chia-Chih by the shoulders and says, "Good! Keep him in your trap. And if you need anything…," Chia-Chih cuts him off asking, "You think I have him in a trap? Between my legs, maybe? You think he can't smell the spy in me when he opens my legs? Who do you think he is?" As Old Wu listens nervously, Chia-Chih continues:

> He knows better than you how to act the part. He not only gets inside me, but he worms his way into my heart. I take him in like a slave. I play my part loyally, so I too can get inside him. And every time he hurts me until I bleed and scream before he comes, before he feels alive. In the dark only he knows it's all true.

As the scene collapses the border between lust and caution, Chia-Chih's words collapse the border between corporeality and desire as she describes his penetrating her and ejaculating inside her body in the same breath as she tells of his penetrating and seducing her mind and her emotions. Despite Wu's protests for her to stop this description, though, Chia-Chih continues:

> Every time when he finally collapses on me, I think, maybe this is it, maybe this is the moment you'll come, and shoot him, right in the back of the head, and his blood and brains will cover me!

Old Wu has no response for her except to yell at her to shut up.

In the end, of course, Chia-Chih fails to remain loyal as she reveals her deception to Mr. Yee the moment before his assassination, and her failure to perform as a seducing spy results in the torture and execution of all the collaborators. This scene with Old Wu returns to the wounded, broken, and bleeding female and male bodies in transnational Chinese cinema. Chia-Chih's retort to Old Wu (and Yu-min) directly links the carnal and psychological in relation to the ruination of bodies, as she draws together the images of her bleeding and screaming, Mr. Yee ejaculating and collapsing, then him bleeding (recalling Tsao's assassination), and his exploded brains covering her. Although described in dialogue rather than depicted in image, this scene brings together much of the enactment of bodily ruination of these films.

In a manner, this collapsing of borders is brought most to light in *Uncle Boonmee Who Can Recall His Past Lives* (*Lung Bunmi Raluek Chat*, 2010), produced and directed by Apichatpong Weerasethakul. This film ruins the divide between the physical and the spiritual, the material and the immaterial, the immanent and the transcendent, the fantastic and the real, especially with regard to bodies. Uncle Boonmee is a widower who is dying and returns to northeast Thailand to prepare for death. He is joined in his preparations by his sister-in-law Jen, his cousin Tong, his Laotian medical assistant Jaai, the ghost of his deceased wife, Huay, who has returned to help him through the transition from life to death, and an apelike being who is Boonsong, his long-lost son who has become a monkey spirit. Halfway through, the film cuts to a fantasy about a princess who wears a veil to cover her facial scars. Beside a tidal pool, she has an intimate conversation with one of her footmen. They appear to be in love. Staring into the pool, she sees her face transform, for a moment, so the scars are smoothed over. Then, she wades into the tidal pool and offers herself to a catfish, with which she shares in a sexual epiphany. Boonmee recalls his past lives to consider how he has gotten to this point and what it may hold for the future. In the end, the film ruins all the boundaries that would arrange and maintain any rootedness through its ruined gaze between and among fantasy and reality, corporeality and desire, or past, present, and future.

It Was No More Than So Many Pieces of Stone

As Olivia Khoo asserts in her analysis of Wong Kar-wai's *In the Mood for Love*, the body in recent Hong Kong art cinema "can no longer be regarded as a dependable marker of identity, whole or fully present, since Chineseness also appears today in increasingly fragmented forms tied to diasporic experience."[20] The current situation, according to Khoo, where attempts to essentialize or reify this disappearing or phantasmagoric body prove to erase it further, calls for a reconceptualization "of an ethical diasporic Chinese viewing position out of the impossibility of rooted Chineseness."[21] Like Uncle Boonmee who recalls his memories as he passes away, the Princess who fantasizes her corporeal integrity in her reflection, and the other ruined bodies/bodies in ruins in transnational Chinese cinema, the attempt to look backward fails because there never was a single, unified Chinese body to look toward as fantasies of "Chineseness" were always already imprinted onto the body and reflected off the body. The attempt to arrange and maintain the Chinese body has failed always, and that failure is what opens the possibility of a future without guarantee.

Rather than further trapping or fixing the image of the Chinese body (and especially the diasporic Chinese body) by making it permanently "visible" and putting it on display once and for all, Khoo argues that we should rethink the ethics of neighbors and locality in regard to these films and the spectral bodies they imagine. Following Khoo, I argue that the focus of transnational Chinese cinemas on ruined bodies/bodies in ruins further provokes a reconceptualization of the relation between film ethics and film bodies in regard to the ambivalent signification of ruins, secrets whispered into crumbling stones, and a temporary reflection in a pool. While ruins mark collapse, destruction, decay, and bankruptcy, they also mark experience, history, memory, culture, preservation, heritage, and inheritance. Thus, rethinking ruined bodies/bodies in ruins foregrounds a rethinking of the wounds, scars, and blood at the origins of diasporic bodies, refocusing diasporic ethics away from origins in times and places toward acts of leaving or being forced to leave: an ethics of ruined bodies relating in transit through ruins. Ruins are, among a great many other things, sumptuous and desolate, palaces and tombs. They are excessive and lacking, empty and overflowing. And they turn us toward the future as they mark the present and the past. In closing, let us turn back to the last lines of Bei Dao's story, "In the Ruins," with which we began:

> Night fell. In the darkness, the outline of the ruins could still be distinguished clearly. He sat on a piece of stone for a long time, then stood up and silently went away.
>
> The noose swayed in the wind.[22]

This hanging rope, too, remains a ruin for the future.

Notes

1. Olivia Khoo, "Love in Ruins: Spectral Bodies in Wong Kar-wai's *In the Mood for Love*," in *Embodied Modernities: Corporeality, Representation, and Chinese Cultures*, eds. Fran Martin and Larissa Heinrich (Honolulu: University of Hawai'i Press, 2006), 239.
2. Khoo, "Love in Ruins," 244–45.
3. Bei Dao, *Waves: Stories*, ed. and intro. Bonnie S. McDougall, trans. Bonnie S. McDougall and Susette Ternent Cooke (New York: New Directions Books, 1990), 6.
4. Christopher Pavsek, *The Utopia of Film: Cinema and Its Futures in Godard, Kluge, and Tahimik* (New York: Columbia University Press, 2013).
5. William Schaefer, "Shadow Photographs, Ruins, and Shanghai's Projected Past," *PMLA* 122, no. 1 (2007): 124–34.
6. Khoo, "Love in Ruins," 242.
7. These ruinous gestures are neither avant-garde nor modernist techniques, ideologically and aesthetically motivated practices set on disrupting the cinematic apparatus and its realist illusion. Such revolutionary techniques often ultimately rely on returning to the truth of the illusion and are rooted in deploying "what really happened." The ruined gaze remains ruined without return.
8. Chris Berry, "Stellar Transit: Bruce Lee's Body or Chinese Masculinity in a Transnational Frame," in *Embodied Modernities: Corporeality, Representation, and Chinese Cultures*, eds. Fran Martin and Larissa Heinrich (Honolulu: University of Hawai'i Press, 2006), 218.
9. Berry, "Stellar Transit," 226.
10. Sharon L. Snyder and David T. Mitchell, *Cultural Locations of Disability* (Chicago: University of Chicago Press, 2006), 160.
11. Ibid., 163.
12. Ibid., 164.
13. Many critics have noted that in the late 1960s and early 1970s Bruce Lee was the most established martial arts actor and fight choreographer in film and television. It was assumed by many that Lee would get the lead role in *Kung Fu* (reportedly a series idea he had pitched to the studios). To the shock of

many in martial arts and entertainment, the role was given to the much lesser known David Carradine. See, for example, Paul Bowman, *Beyond Bruce Lee: Chasing the Dragon through Film, Philosophy, and Popular Culture* (New York: Columbia University Press, 1993) for a racial and ethnic reading of why Lee was not cast as the lead in the series.
14 Rey Chow, *Primitive Passions: Visuality, Sexuality, Ethnography, and Contemporary Chinese Cinema* (New York: Columbia University Press, 1995), 171.
15 Su Tong, *Raise the Red Lantern: Three Novellas*, trans. Michael Duke (London: Scribner, 1993), 98.
16 Song Hwee Lim, *Celluloid Comrades: Representations of Male Homosexuality in Contemporary Chinese Cinemas* (Honolulu: University of Hawai'i Press, 2006), 69.
17 Song, *Celluloid Comrades*, 88.
18 Jami Proctor-Xu, "Sites of Transformation: The Body and Ruins in Zhang Yang's *Shower*," in *Embodied Modernities: Corporeality, Representation, and Chinese Cultures*, eds. Fran Martin and Larissa Heinrich (Honolulu: University of Hawai'i Press, 2006), 174.
19 Michel Foucault, *Discipline and Punish: The Birth of the Prison*, trans. Alan Sheridan (New York: Vintage, 1977).
20 Khoo, "Love in Ruins," 235.
21 Ibid.
22 Bei Dao, *Waves*, 156.

CHAPTER 2

Transnational Cinema as a Matter of Address: Considering Eros and Embodiment in Wong Kar-wai's "The Hand"
Sim Jiaying

Abstract

This chapter situates Wong Kar-wai's "The Hand" in its transnational context through its relation to the American and Italian segments of the anthology film *Eros* (2004). The fact that "The Hand" was meant to appeal to an international audience highlights the importance of considering the roles of "modes of spectator address" and reception when considering conceptualizations of transnational cinema. In discussing these conceptualizations, this chapter will engage in an examination of the cinematic address and film style of "The Hand" to suggest we reconsider the eros and embodiment of *Eros* in terms of more complex understandings of desire and corporeality. In the end, this reconsideration allows for a rethinking of "transnational cinema" beyond simple categories of geography, signification, and representation.

Keywords: Transnational Cinema, Affect, Sensuous, Embodiment, Spectatorship, Modes of Address, Wong Kar-wai, *Eros*

Wong Kar-wai is one of the best-known directors of contemporary film, and since *As Tears Go By* (1988) first gained international recognition at the 1989 Cannes Film Festival, his name has become synonymous with Hong Kong Chinese cinema. It seems simple and straightforward enough to categorize Wong as a Hong Kong Chinese director, especially when one considers his nationality, the production locations of his films, and the language used in his films. However, Wong's body of work is almost always regarded as more than just representations or products of Hong Kong Chinese cinema as his films exceed most conceptions of "national cinema." Yet, while critics continue to label Wong's work as "transnational," they fail to explain the exact justification for that label. I argue that despite the seemingly national aspects to Wong's films, the tensions produced by Wong's cinematic mode of address, the spectatorial experience they produce, and the spectator reception they elicit, reshape our conceptions of what makes his films transnational.

To illustrate this claim, I attend to Wong's short film, "The Hand," the first of a three-episode feature film, *Eros*, which also involved American director Steven Soderbergh ("Equilibrium") and Italian director Michelangelo Antonioni ("The Dangerous Thread of Things"). *Eros* was first screened at the Venice International Film Festival in September 2004. From the start, then, *Eros*'s international production and intended international distribution and exhibition positioned it outside national film paradigms. The film was never marketed as a product of national cinema, nor was its intended viewership imagined to be restricted to any one of these three nations.[1] Already, we can follow what Marsha Kinder contends in *Blood Cinema: The Reconstruction of National Identity in Spain* (1993)—that there are limitations in addressing cinema based on the theorizations of "national cinema" because "[cinema] must be read against the local/global interface."[2] Likewise, in light of *Eros*'s international collaboration and its intended international audience, we might regard "The Hand" with respect to Andrew Higson's discussion of national cinema in his contribution to *Cinema and Nation* (2000). Higson asserts that the nuances of cultural and economic cinematic productions cannot be reduced easily to national parameters and suggests that "the contingent communities that cinema imagines are much more likely to be either local or transnational than national."[3] Read against the local/global interface and accepted national parameters then, we might easily classify *Eros* as global or "international." But can we classify it, or its three segments, as "transnational," and what does it teach us about the category "transnational" if we do so? What does it mean to think transnationally about cinema? What does it mean to think transnationally about *Eros* and "The Hand"?

The fact that the "The Hand" occupies a unique position as both an individual work that contributes directly to the sphere of Hong Kong Chinese cinema *and* as part of an international omnibus feature film offers an interesting way of thinking not only about which films count as "transnational" but also about what transnational cinema does and can do. Considering "The Hand" and *Eros* as examples of transnational cinema challenges extant theories of transnational cinema. Current theories of transnationalism tend to define a transnational film by its production contexts, yet they fail to consider the transnational aspect of cinematic experience itself. When audiences experience a film, they are moved from the context of their own experience into the diegesis of the filmic world, yet they never fully leave behind the embodied experience tied to their own histories and social contexts. This is the transnational affect of "The Hand" that resonates throughout the film segment and one that is encouraged through the film's sensuous mode of address. Examining closely this affective aspect of "The Hand" exposes gaps within current transnational theories based on production.

Considering "The Hand" and *Eros* in terms of their production, distribution, exhibition, and reception means considering three directors from three nations producing three film segments (or chapters) to make up a larger film—which we may or may not consider a totality—that debuted on an international platform. However, thinking of how to identify "The Hand" as either local or global, as either national or international, misses the point of its mode of address. Rather than thinking in terms of the film's identity, of what it *is*, we might think of the film in terms of what it *does*. The placement of "The Hand" within *Eros* builds a tension that holds the local in relation to the global, the national in relation to the international. Thus, it maintains a heterogeneity that resists flattening or forgetting. This maintenance is one of the things the film's mode of address does, one of the things transnational films do. Furthermore, as technologies change and the film continues to be distributed and exhibited in different ways—especially alongside other films and new media—this tension will continue to increase and change. The complexity of these factors means questions of "transnational" cinema are much more than questions of origins, locations, languages, or settings. Cultural and political nuances across and within national and international boundaries are often neglected within discussions of transnational cinema, especially when the production, distribution, exhibition, and reception of films complicate those local and national boundaries. Such complications are even more exposed when we go beyond the identity of these films and examine their affect.

What we can see, then, is that "The Hand" and *Eros* do not fit neatly into simple models of transnational cinema. "The Hand" is hardly bound by geography, proximity, or a relation between the margins and the center of dominant cinema. Thus, the short film and the larger omnibus into which it fits challenge preexisting models of conceiving transnational cinema, and it is this inability to properly locate "The Hand" within current paradigms that ensures a critical analysis of what "transnationalism" means. Rethinking "The Hand" offers us the chance to rethink our conception of "transnational cinema." What happens when we turn from asking how *Eros* and "The Hand" fit into existing models for transnational (Chinese) cinema to asking how they provide us with a more complex model for transnational (Chinese) cinema? Along the way, through an in-depth analysis of the mode of address and style of "The Hand," in particular, we may also gain a better understanding of the complex relation between cinematic desire and corporeality and a better understanding of "eros" and embodiment in the overall film. In postulating how "The Hand" deals with eros through an embodied cinematic experience, as compared to the other segments of *Eros*, I proffer a new way of understanding transnational cinema through modes of address that function beyond signification and representation.

Regarding filmic modes of address and cinematic experience opens the possibility of responding to films through transgressing and transforming national boundaries while transporting audiences through time and space—regardless or in spite of apparent national boundaries. While there remains value in turning to production backgrounds, financial circumstances, generic concerns, and even regional synergies in describing films as national, international, or transnational, the model advocated here insists on also focusing on the relation between films and audiences. What are the significant thematic, formal, and stylistic elements of "The Hand"? How do they relate to the overall theme, form, and style of *Eros*? What might addressing theme, form, and style bring to discussions of transnational cinema and cinematic experience?

All three segments of *Eros* address the themes of love, sex, and desire—each providing a different look at the complex embodied relations among these themes. However, as each director is known for a different film style and each has a long history of treating these themes in distinct ways, audiences may expect and respond to the modes of address put forth by the segments in the light of their separate forms and styles. Thus, audiences may not necessarily reduce the whole film to one synthetic form or style. Paying attention to the individual modes of address within the segments of *Eros* creates a tension among the segments and between segments and the whole film, though. Therefore, heeding the relation among theme, form, and style becomes a starting point in considering the "transnationalism" of *Eros* and "The Hand" and raises the question of the tension that may be at the heart of the transnational. Each director provides varying perspectives on what eros entails—as in desire and love and sexual yearnings—yet there is something about eros that inevitably escapes a definitive, tangible, and all-encompassing understanding. Audiences cannot easily deduce a stable conception of eros simply by synthesizing it from these three episodes. Attending to the thematic concern of these three films exposes the various modes of address involved in expressing content associated with eros. Thus, the tension among the three segments prompts us to consider how this particular set of filmic discourses on eros, desire, and corporeality seeks to move and affect the audience.

In the second episode, "*Equilibrium*," Soderbergh investigates the idea of eros as a state of mixed consciousness by blurring the lines that distinguish dreams from realities. In this episode, an alarm clock salesman, Nick Penrose (Robert Downey Jr.), visits his psychiatrist, Dr. Hal Pearl (Alan Arkin), to discuss a recurring sexual dream he has about a woman he does not know, only to wake up to the realization that the psychiatric session was merely a dream. Waking from that dream, Penrose is greeted by his wife who turns out to be the mysterious woman from the sexual dream he speaks about in his psychiatrist's office. Furthermore, the audience is left to question even these fantasy/dream/reality boundaries because Penrose sometimes, in his dream, describes events that occur after he appears to awaken. At this point, Penrose's dream seems to have seeped into his reality and his reality into his dream. His conscious and subconscious begin to blur, and the line between reality and dream becomes more indiscernible as the film continues. Or, as Dr. Pearl says toward the end of the film, "It's a nightmare isn't it? We're living in the shadow of the freaking snooze alarm. I mean, where do you go from there?" Rather than providing some form of closure or definitive response, the scene cuts to a medium close-up of Penrose staring blankly to the side as the sound of percussions crescendo in the background. The film ends with a worm's-eye camera view of the building and sky, where repeated scenes of a paper plane being thrown from one of the windows are disjointedly cut together—while the background percussion plays on, unaffected. Here, the film form works to disorient the audience through a series of cyclical repetitions rather than a revelatory conclusion.

Thus, the film's repetitious ending reiterates for the audience the difficulty Nick has experienced in pinpointing (his) sexual desires through psychoanalysis. Nick struggles within "Equilibrium," and the audience struggles while experiencing the segment and while trying to fit the segment into the whole of *Eros*. Here, one may think about this labyrinth of narratives, entangled with one another rather than unraveled neatly and coherently, as one way of expressing the theme of eros. Furthermore, this is only one of the ways in which the film addresses the audience—through the display of a semantic code open to decoding into a knowable narrative that is reiterated through audio-visual means. While this is a functional mode of address and could be considered the primary way that we think about the cinematic experience, the fact is that a model based so emphatically on the decodable and denotable signs of the audio-visual closes the discussion surrounding transnationalism significantly. In fact, this audio or visual model of cinematic engagement is very often thought of as the only way that cinema communicates with the audience, perhaps because it covers over important questions and makes the complexity of the cinematic experience more manageable.

In the third segment, "The Dangerous Thread of Things," Antonioni also approaches the concept of eros through a narrative that ends more than it concludes. Not much happens in the film. It is more interested in complicated characterization than narrative development. A married couple (Christopher Buchholz and Regina Nemni) is having a difficult time of it. Throughout the film they take walks that end in intense fights or they eat in silence at a beachfront diner. At one point, a young woman enters the diner. Eventually, she and the husband sleep together, and their affair threatens the marriage even more. The film ends with crosscutting between scenes of the young woman and the wife dancing naked on the beach separately, at different times. In the final instance of the segment, they look at each other before the screen fades to black. Again, the film makes it difficult for the audience to decipher the trajectories suggested by the ending of the film.

Unlike "Equilibrium," though, "The Dangerous Thread of Things" focuses more on an overt and visual display of embodied eros as the latter episode is filled with many more scenes of nudity and sexuality. Thus, we can think about "Dangerous" as a film that suggests that eros is produced and expressed visually, through sexual imagery rather than through narrative development and conclusive statements about the nature of eros. This episode brings to the fore how films can function prior to meaning, how they can address an audience without making explicit meaning their guiding principle. The episode's mode of address exposes the cinematic medium as experiential, where the storyline and lack of comprehension or coherence are secondary to the embodied experience of affect. Here, when we take into account modes of address in theorizing transnationalism we see affect (as with tension earlier) as an element that takes precedence over narrative and meaning making.

Finally, the first episode of *Eros* is Wong Kar-wai's "The Hand," which some critics have called the most "readable and watchable" segment of the feature film.[4] Unlike the other segments, the narrative of "The Hand" is more direct. A young apprentice tailor, Zhang (Chang Chen), makes dresses for a sex worker, Miss Hua (Gong Li). While Zhang waits in the living room to see Miss Hua for the first time, he is aroused by the sounds of Miss Hua having sexual intercourse. Miss Hua realizes this when she sees him and orders him to take off his pants. She then clinically and mechanically masturbates him, explaining, "How can a tailor make dresses for a woman if he has not touched nor been touched by a woman? If he remembers her hand, he will make good dresses in the future."[5] Zhang eventually becomes a professional tailor, while Miss Hua continues in her profession. They meet consistently over the years for Miss Hua's dress fittings and alterations, but Zhang's desire for her is never requited. Eventually, Miss Hua's health and reputation slowly decline because of her tuberculosis. In the final days of her life, the two repeat their first encounter but with more passion than before as they share each other's touch and share in each other's desire.

Unlike the other episodes, "The Hand" does have a coherent narrative. Yet, it would be incorrect to suggest that Wong's film is the most effective exploration of eros in the omnibus film simply because of its coherence. "The Hand" is effective, in part, because of its narrative development but also because its mode of address highlights that very cinematic potential to tell a coherent story as well as to evoke cinematic functions that are not dependent upon narrative. The film's mode of address emphasizes cinematic narrativity, affectivity, and sensibility. It is these last two that are often neglected in studies of transnational cinema and the analysis of these two that shows us how a close look at "The Hand" might change the ways we theorize transnational cinema as a whole. By addressing narrativity, affectivity, and sensibility, "The Hand" demonstrates (tells, shows, emotes, and "touches on") how eroticism functions through the cinematic experience. "The Hand" demonstrates this function through audio-visual-haptic mediation; it addresses its audience as an embodied audience, not just disembodied eyes or ears in a cinema. In a way, it touches its audience and is touching in its affect. This act of touching the audience is maintained throughout "The Hand" and remains with the audience during the omnibus because the resonances of its narrativity, affectivity, and sensibility function as phantoms through their absence in the next two segments. Thus, the audience is haunted by the way "The Hand" balances these three aspects seamlessly, so much so that the cinematic experience permeates the next two films. In these moments, when the modes of address of the different segments are compared and considered, "The Hand" has already functioned transnationally.

The story begins in the narrative present with Zhang asking after Miss Hua's health. Miss Hua tells Zhang not to visit anymore as her illness is contagious. Then, she asks, "Do you still remember my hand?" and they remember the first time they met. The scene cuts to the stairway that leads to Miss Hua's house where the events of their first encounter occur. The audience is now in the narrative past. The rest of the film unfolds through this flashback, only returning to the narrative present at the end. Only five minutes into the story, the audience knows that Miss Hua and Zhang have a history together and that Miss Hua is suffering from a contagious disease.

This is a story that can be told in a few minutes. Yet the point is that "The Hand" reiterates the notion that eros emerges not so much through the content and complexity of a discernible story line but through an embodied cinematic experience. Through this embodiment, cinematic experience entails an aesthetic and affective aspect that involves and affects the audience not only by relying on making meaning but also on making sense. Prior to signification, the audience is already affected by the film, whether or not we are consciously aware of the exact and distinguishable factors contributing to such sensuous affects. For example, the long establishing shots that open the film do not do much in establishing the context of the film but provoke the audience to experience these scenes as they are. They immerse the audience into the film. This immersion is essential to the embodied experience of Wong's transnational mode of address. The tangible, signifying facts of the narrative do one thing while the intangible, non-signifying aspects of the mode of address do something else. While we decode and understand the tangible plot points, we also feel the emotional and psychological affects of the characters. We experience their eros.

During a pivotal scene where Miss Hua asks Zhang if he needs to take her measurements again—since she has lost weight—Zhang replies calmly that he remembers her figure; all he needs to do is to use his hands and he will know. As the soundtrack plays, we see a medium close-up shot of the two protagonists' side profiles, with Zhang standing behind Miss Hua. Slowly and steadily, he traces his hands along Miss Hua's shoulders and down the sides of her arms. The camera lingers again on the two characters while his hands move out of sight. Miss Hua takes a quiet breath and closes her eyes. Desire is made palpable. The camera cuts to another medium close-up of Zhang's hands near Miss Hua's waist. The dark silhouettes of Zhang's tuxedo and Miss Hua's dark cultured "cheongsam" cover much of the scene. Zhang's hands can barely be seen. The camera cuts again to a medium close-up of Miss Hua's hands on her waist while she holds tightly onto Zhang's hands. We feel when the characters touch.

Eventually, Zhang's hands embrace Miss Hua's hands passionately as they hug without exchanging a single word. Against the background instrumental music, we hear slow, deep breathing. The camera then alternates between shots of Miss Hua and Zhang hugging, solemn and emotional, and shots of their hands entwined passionately before cutting to a long shot of their back views in the dimly lit motel room. Then, the camera pans right, and the screen fades to black. Even in this scene where there appears to be a moment of development between Zhang and Miss Hua, Wong's mise-en-scène remains cramped and dark, and it is difficult for us to discern exact character movements and gestures. While it is convenient to think that our eyes function like hands in this scene, tracing the touch between Zhang and Miss Hua, the fact is that our eyes are blocked by the obscuring mise-en-scène, not given complete access to the images on screen. As such, eros here is not expressed only through signification but through the cinematic medium—the camera that manages to tease out an embodied haptics of eros: the unverbalized senses and sensations of touch, which constitute the significant while non-signifiable and non-representable aspects of eros. Through mise-en-scène and camera movement, Wong coaxes us into feeling what we cannot see.

This moment in "The Hand" aligns with what Laura Marks contends in *The Skin of the Film*. In that text, Marks addresses the complexity of the cinematic experience when she explains "film signifies through its materiality, through a contact between perceiver and object represented... the way vision itself can be tactile, as though one were touching a film with one's eyes."[6] For Marks, film presents a contact point with the audience through the cinematic medium's embodiment, which functions like a kind of skin. As such, as the film reaches out to the audience, the audience is encouraged to touch the film with their eyes, rather than to just *see* a film. This is what she deems "haptic visuality."[7] In this way, the film functions like a corporeal *body of cinema* because the audience, which may be regarded as the *body at cinema*, is provoked to interact with it through touching, instead of only through seeing and hearing. It is through this embodied haptics of cinema (haptics of eros) that the audience is affected by the film, and especially by the affectivity of the mode of address performed by our concern over the relationship between Zhang and Miss Hua, the *bodies in cinema*.

This scene demonstrates a telling example of the complex embodiment of what Marks describes as "haptic cinema" in *Touch: Sensuous Theory and Multisensory Media*. For Marks,

> Haptic cinema appeals to a viewer who perceives with all the senses. It involves thinking with your skin, or giving as much significance to the physical presence of an other as to the mental operations of symbolization.... Haptic cinema, by appearing to us as an object with which we interact rather than an illusion into which we enter, calls on this sort of embodied intelligence.[8]

This mode of interaction relates to film's materiality that goes beyond what is apprehensible to the eyes and ears. In other words, Marks posits that it is important for the audience to acknowledge the "physical presence" of the film through more proximal senses, such as touching, whether or not this interaction results in comprehending the film through intellectual meaning-making linked to seeing or hearing.

The proximity between film and audience ensures that the film no longer functions as a text that the audience deciphers or decodes or works to contain within a fixed understanding. Rather, the filmic body affects the embodied audience as the embodied audience affects the filmic body, creating potentialities for cinematic experiences surpassing interactions solely based on the film being read as a fixed text. Haptic visuality and embodied modes of address make possible more complex cinematic experiences because the audience becomes a variable element in the experience rather than just a master over a text. It is precisely this productive rather than reductive mode of address that opens up transnational cinema because our cinematic experience of a film is not dependent only on where the film was made but also based on the differences—social, economic, political, historical, national, gender, and such—of each audience member who adds to the singularity of each experience. In other words, every audience takes from and adds to the filmic experience and not only by decoding or deciphering it in different ways or with divergent biases or points of view. Each specific film experience is made different by the different audience members who experience also the intangible, non-signifying aspects in specific ways. Addressing these differences is at the heart of the transnational mode of address.

As much as we may aim to objectively describe the cinematic techniques and styles used in the films, our interpretations remain subjective because they are based in the personal and subjective experiences we have with subsequent interactions with these films. However, in sharing this cinematic experience with the filmic body, a transnational exchange has already occurred and been produced. This is not to say that this exchange is one that can easily be broken down and analyzed. With regard to "The Hand," it is also interesting to note that despite having sexual and erotic content in the narrative, no naked bodies are shown on screen, nor are the sexual encounters between Zhang and Miss Hua, or Miss Hua and her clients, ever explicitly shown on screen. Without needing to show the audience these provocative scenes, or without the audience needing to see to know, Wong's film evokes a sensuousness and eroticism that is not dependent on purely visual indicators. Thus, the film affects our senses beyond our need or ability to comprehend. In other words, "The Hand" is able to express eros as a mode of address through its mode of address.

This conceptualization of the cinematic experience does not suggest the camera functions as a disembodied eye, which imitates the sense of touch through a primarily visual cinematic experience. To think of the camera and the cinematic experience in this way is to remain in the realm of the disembodied and omniscient kino-eye with Dziga Vertov or abide by traditional apparatus theory, which argues the subject at the cinema becomes little more than an isolated, passive, disembodied eye interpolated by the totalizing cinematic interaction. In acknowledging and celebrating the potential that images have in their appeal to the body regardless of any other appeal through the audio-visual tracks alone and knowing eye during cinematic experiences, "The Hand" brings to the fore an embodied experience of eros and an experience of eros as embodiment. "The Hand" makes plain how cinematic eros functions through embodiment. Just as our embodied experience with the real world can hardly be broken down to minute analysis, our cinematic experiences are embodied in ways that go beyond purely decodable audio-visual modes of address. The cinematic medium by way of its direct appeal to our desire and corporeality translates the lived experiences of eros through an embodied spectatorship to the extent that the audience no longer needs to understand in order to experience cinematically, because comprehension of the audio or visual cues on screen is not a prerequisite to being moved or affected by a film.

Marks reminds us of the connection between eroticism and haptic cinema:

> What is erotic about haptic visuality, then, may be described as respect for otherness, and concomitant loss of self in the presence of the other. The giving-over to the other that characterizes haptic visuality as an elastic, dynamic movement, not the rigid all-or-nothing switch between an illusion of self-sufficiency and a realization of absolute lack.[9]

In other words, haptic visuality is a mode of address that appeals to the audience, reaches out to it, in an interaction that is no longer dependent upon imposing (or receiving) a totalizing interpretation of the filmic experience. Yet, this mode does not demand the audience surrender completely to the point where the film overwhelms and overpowers each singular cinematic experience. Through the porousness of this mutual regard, eroticism is evoked because the "exchange" between film and audience is no longer just a simplified "either/or" interaction. It is no longer a direct exchange. In fact, it is this notion of alterity that ensures the address and reception of "The Hand" remains transcultural and transnational because we can never completely lose ourselves in a cinematic experience we can never claim solely as our own. We are addressed and affected but never interpolated or sutured. At the same time, though, we are also never wholly separate from that address and affect.

The affect of eroticism is not evoked only through the camera movement, but also through its supposed lack of movement as well. Just as in Wong's other works, such as the earlier *In the Mood for Love* (2000) or the recent *The Grandmaster* (2013), the camera in "The Hand" lingers through scenes of long takes and deep focuses almost without purpose. In the opening establishing shots, we see three scenes. The first is that of a dimly lit motel signboard with a small street lamp barely illuminating the scene. The shadow and dark silhouettes of the door create a visual spatiality that allows little to be seen and less to be discerned. The second shot is of a cramped and dimly lit corridor inside what seems to be the motel. The mise-en-scène is organized but cluttered because the dim lights remain the most obvious and identifiable objects on screen while the doors overlap onto one another as we struggle to distinguish the details in this setting. And finally, the third shot is a medium close-up of a solemn-looking Zhang in the narrative present, talking to Miss Hua.

Shots that resemble these in the opening are repeated throughout the film—be they of the living room where Zhang waits to see Miss Hua, the stairway that leads up to Miss Hua's place, or views of the characters' backs as they stand against different backdrops. These scenes where nothing or nothing much appears to be happening evoke the experience of freeze frames. They are not freeze frames, yet the experience of such mise-en-scène affects us in a way that mimics our experience of freeze frames. This lack of visible movement on screen does not function as a cue to think about these scenes as still images or photographs in a slideshow that Wong utilizes to supplement the story he tells. Jean Ma explains this general misconception about the freeze frame in her essay "Photography's Absent Times." The freeze frame is often misunderstood as "the image of a single photogram stalled in the projector, when in actuality the device consists of a repeated series of identical frames."[10] The repetition of a still image thus creates an illusion of stasis through movement. Freeze frames invite us to rethink the notion of stillness as the absence of movement in cinema.

With freeze frames, even when we feel we are experiencing a still image, the fact remains that the scene is always still in motion, whether or not its movements are visible to the eye. As such, we do not regard these freeze frames or apparent still images as fixed entities removed from a linear progression. In fact, the paradoxical nature of the freeze frame underscores how cinematic experiences are characterized by motion and fluidity, whether or not these aspects are perceivable (visible) at first glance. The illusion of stillness and lack of noticeable changes in freeze frames in general and in these scenes in particular highlight how cinema does not function solely (or perhaps even mostly) through details that are visible to our eyes but through imperceptible elements as well. In this manner, it is important to consider how freeze frames, long takes, and deep focus emphasize the sensuous quality of cinema without necessarily contributing any signification to the storyline. They highlight the sensuous function of all the affective aspects of the cinema. Therefore, they call attention to the overall embodied aspects of the cinematic experience. It is often easy to mistake the fact that just because we are unable to see any differences and movements in the filmic image, nothing is happening. In fact, freeze frames or long takes that appear to show us nothing at all do precisely the opposite.

Through the sensual experience of the freeze frames we experience an embodied passage of time. We sense the passage of time and the longing marked by that passage. The experience of the passage of time where little happens between Zhang and Miss Hua is the point. We feel their time together. We do not identify with Zhang because these long takes and freeze frames of empty corridors, living rooms, and doorways are not aligned with his look. Rather, we are already involved in this cinematic experience of time. We wait with the characters for something to happen in these scenes and we are made to wait over and over again. We may not even become aware that we wait, expectantly, but we experience it perhaps even more as we are unaware we are waiting for something to happen. In fact, prior to realizing that time has passed in the filmic world or prior to realizing that the characters now wear their hair differently or that Zhang has made more dresses, we are already affected by the passage of time. This cinematic mode of address functions without demanding we understand or know what we experience. Before cinema makes meaning, it makes sense. Only when one attends to the affective modes of address that act on one's senses does one begin to attend to the dynamism of our cinematic experiences. Freeze frames draw the audience's attention to the passage of time that may or may not remind one of the inevitability of time on Zhang's unrequited desire toward Miss Hua, or the inevitability of time that takes a toll on both Miss Hua's health and reputation. Our experience of the film time of "The Hand" makes palpable Zhang's desire, makes corporeal the eros of the film. Even though we may not understand their relationship, experiencing the film becomes an embodied experience of eros.

By slowing down our experience of eros, by lengthening it, the mode of address of "The Hand" prompts us to regard the film in terms of its otherness. It resists us imposing our point of view on it. This is further underscored by the importance of the soundtrack in this particular cinematic experience. The music heightens our sense of the drama here, and of the feelings between the characters, even after we know nothing has changed between the characters. Zhang will continue to desire Miss Hua. Miss Hua will continue to maintain her distance. The music does not indicate a narrative shift or present a turning point in the story. Zhang's desire remains unrequited, and we may feel his pain even more without knowing anything more about it. This affective quality is what Jeremy Gilbert describes in his essay on the function of film music. In it, Gilbert asserts,

> Music has *physical effects* which can be identified, described and discussed but which are not the same thing as it having *meanings*, and any attempt to understand how music works in culture must ... be able to say something about those effects without trying to collapse them into meanings.[11]

This echoes Deleuze's contention in *What Is Philosophy?* that art is "a pure being of sensation."[12] Unlike philosophical concepts, which are conceptually understood, aesthetic objects (and their aspects) are aesthetically experienced. Film and film music cannot be understood only through semiotics or structural approaches because it makes an impression before cognitive comprehension takes place. At a primary level, music expresses itself through the embodied experience of the audience and that of the film and cinema. The body at cinema comes into contact with the musical and sensuous body of cinema to create new affective environments.

In this way, music is affective not because it functions as an effective signification. Rather, its propensity to affect and to be affected derives from the way music exists as sensation. Music functions before it means anything. Perhaps, critics who assert that "The Hand" is the most readable and, therefore, the most enjoyable of the three segments of *Eros* are missing the point. It is not about the readability of the film, if we take readability to imply understanding. Eros, as in love and passion and desire, cannot be restricted to a set of criteria or factors based on signification. It is a mood, a sense, an impression that is experienced through the body prior to an intellectual grasp of the experience.

Through the aforementioned analysis of the modes of address that communicate and express eros in the film, I highlight how eros relies both on a sensorial and affective experience as much as narratability. In fact, the affectiveness of eros as produced through film ensures the cinematic experience transcends the boundaries of nations and culture where the film was made. As such, in attending to the modes of address in relation to the cinematic content of eros, the failure and limitations of a solely audio-visual mode of cinematic experience are brought to the fore. In other words, whether or not the audience understands the (cultural) connotations of Miss Hua's dresses, which are actually traditional mandarin gowns or "cheong sam," is also beside the point.[13] The fact is that the audience, in recognizing that there is an unknowable aspect to the film—be it the significance behind Miss Hua's need to have her dresses tailored, if any, or in the significance of Miss Hua and Zhang's relationship, if any—and acknowledging a difference and distance that exist between spectator and film, one is already using an approach that is as transnational as it is ethical.

This is because the audience is unable to maintain a stable or definitive reading of the film, as its modes of address destabilize any fixed grasp on the film before it can be firmly established. Even as the audience is simultaneously offered a space to regard the film in proximity through an embodied mode of address, while not being able to assume to know the film in its entirety, audience and film acknowledge each other's differences across this plane of cinematic experience that is pointedly ethical in the refusal for meaning to be imposed on either one party (film or audience). In brief, audience and film do not mesh up into a unified totality that should be *the* cinematic experience of "The Hand." Rather, its transnational quality is evoked through a singular cinematic experience that takes into account the uniqueness of the audience's social, historical, and cultural context, in relation to that of the film's modes of address that surpass identifiable and constant production factors.

Each time we revisit this film we come to it differently—be it through time passing or otherwise—and this difference is transferred to the way we experience the film. However, such vicissitudes are usually ignored when we label films "national," "international," or "transnational" based on a fixed set of prefilmic categories or a fixed meaning or interpretation where the social or historical context of the audience does not matter. In attending to modes of address, transnationalism is theorized by taking into account the changeability of the audience's cinematic experience of an ever-changing film, thus ensuring that the meaning of transnationalism is constantly revisited and reconstructed with each "viewing" or experience of a film, especially a film such as "The Hand."

As such, it is not to say that the other two segments in the film are not considered transnational because through a focus on modes of address, the audience may also be transferred across national boundaries or parameters in a way that bridges the differences between the audience's and film's positions, but not one that erases heterogeneity in place of homogeneity. Undoubtedly, they can be considered transnational. However, unlike the other two films, while Wong's segment may be considered transnational, it also stands as a film that allows us to contemplate more fully the potentially political or ethical implications of such transnational modes of address that are effected through filmic aspects other than purely audio-visual ones. In this way, when we attend to film based on its modes of address, we more fully address the question of "What does transnational film do?" rather than "What is a transnational film?" The failure of the audio-visual mode of address in this particular short film compels the audience to turn to other modes of address that do not rely only on meaning-making. These methods affect us by transforming and transmuting the distance between film and audience to include a dynamic potentiality of how we regard a film and the conceptualization of transnationalism.

Notes

1. See film review, Nicholas Y. B. Wong, "Film Reviews: 'The Hand,'" *Scope* 30, no. 4 (August 2013), http://www.scope.nottingham.ac.uk/filmreview.php?issue=4&id=123.
2. Marsha Kinder, *Blood Cinema: The Reconstruction of National Identity in Spain* (Berkeley: University of California Press, 1993), 7.
3. Andrew Higson, "The Limiting Imagination of National Cinema," in *Cinema and Nation*, ed. M. Hjort and S. MacKenzie (New York: Routledge, 2000), 63.
4. Wong, "Film Reviews: 'The Hand.'"
5. Unless otherwise indicated, translations of the film are those of the author.
6. Laura Marks, *The Skin of the Film: Intercultural Cinema, Embodiment, and the Senses* (Durham, NC: Duke University Press, 2000), xi.
7. Ibid.
8. Laura Marks, *Touch: Sensuous Theory and Multisensory Media* (Minneapolis: University of Minnesota Press, 2002).
9. Ibid., 18.
10. Jean Ma, "Photography's Absent Times," in *Melancholy Drift: Marking Time in Chinese Cinema* (Hong Kong: Hong Kong University Press, 2002), 67.
11. Jeremy Gilbert, "Signifying Nothing: 'Culture,' 'Discourse' and the Sociality of Affect," *Culture Machine* 6 (2004), http://www.culturemachine.net/index.php/cm/article/view/8/7.
12. Gilles Deleuze, *What Is Philosophy?*, ed. Lawrence D. Kritzman, trans. Hugh Tomlinson and Graham Burchell (New York: Columbia University Press, 1994), 9.
13. In the 1920s, the "cheong sam" was a figure-hugging dress that was worn by upper-class women in Shanghai. With this knowledge, the audience might suggest a reading of Miss Hua as maintaining a sense social class by means of having her dresses tailored, and always wearing these "cheong sam." Or perhaps draw upon a supposedly more conclusive reading or analysis of the film.

CHAPTER 3

Female Chinese Stars on Screen:
Desiring the Bodies of Ruan Lingyu and Linda Lin Dai
Mary Mazzilli

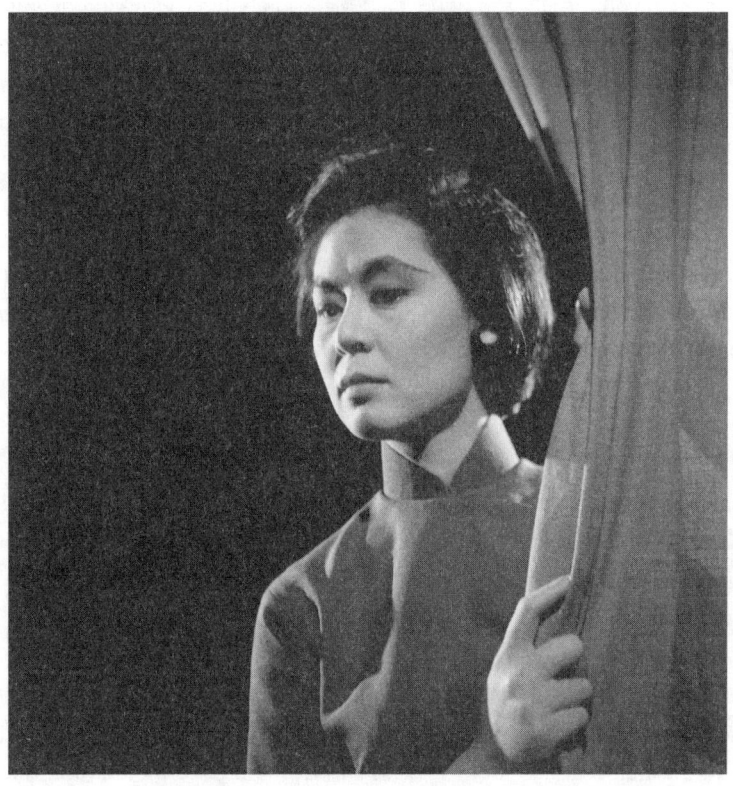

Picture 3.1 Still from Love Without End 不了情

Abstract

According to some scholars, the male directorial gaze objectifies the female body on screen and female subjectivity for its own pleasure, coercing them to perform and display this very desire. Chinese feminist critic Dai Jinhua suggests that such coercion is not only inevitable but also necessary for the production of cinematic pleasure. By looking more closely at the embodiment of particular Chinese female film stars on and off screen, however, this chapter argues that we can discover that Chinese female film stars are often not passive, not victims of the external directorial gaze, but display their effort to find their own place in society and their own (using Nietzsche's words) "will to power." This study discusses female Chinese stardom by considering corporeality and desire, and Merleau-Ponty's conception of visibility, which articulates Chinese female stars' embodied cinematic experience. The Chinese female stars under scrutiny are Ruan Lingyu (1910-1934), from the 1930s Golden era of Chinese cinema, and Linda Lin Dai (1930-1964), from 1950s Hong Kong cinema.

Key Words: Desire, Failure, Merleau-Ponty, Hong Kong, Female Stars

In her essay, "Woman, Demon, Human: The Plight of a Woman" ("*Ren, Gui, Qing: yige nurende kunjing*") prominent Chinese feminist and cinema critic Dai Jinhua discusses the 1987 film *Woman, Demon, Human (Ren, gui, qing)* by Chinese female director Huang Shuqin. In this essay considering issues of gender, sexuality, and embodiment in the film, Dai writes,

> The subject of the woman seems mainly one of silence, one that is always metaphoric [....] That is the image of a naked woman that had rushed through the dreams of men, silent, she never existed, she never will ..."[1]

As Tani Barlow explains, "When, in Dai's view, women are no longer an object of male desire, they cannot represent themselves as Other, and consequentially they cannot fulfill their emancipatory potential as the sign of gendered difference."[2]

In discussing representations of female bodies on screen, film scholars often describe how the male directorial gaze objectifies the female body and female subjectivity for its own pleasure. According to such critics,[3] established visual codes and narrative norms not only subject the female body on screen to the desire of the camera gaze but also coerce it to perform and display this very desire. This circular dynamic, in the words of Dai Jinhua, is not only inevitable but also necessary for the production of cinematic pleasure. Against this established paradigm and Dai's endorsement of it, this chapter argues that by looking more closely at the embodiment of particular female Chinese film stars on screen, we can discover that they are often not passive, not victims of the external directorial gaze, but also enact their own place in society and their own (using Nietzsche's words) "will to power." In this study, female Chinese stardom is aligned with the confluence of corporeality and desire, which, following Merleau-Ponty's conception of visibility, articulates Chinese female stars' embodied cinematic experience. This is particularly true in the case of two female stars: Ruan Lingyu (1910-1934), from the 1930s Golden era of Chinese cinema, and Linda Lin Dai (1930-1964), from the Golden era of the Mandarin musical.

According to Zhang Zhen, already in the early 20th century in China cinema had attracted a female spectatorship.[4] In his book *Amorous History*, he refers to an approach to cinema that inserts "the historical female spectator qua modern consumer into cultural and social history of cinema" and "recognizes women's participation in cinema as an alternative public sphere."[5] The element of a growing female spectatorship and the recognition of women's roles as consumers of the cinematic experience make the role of female stars all the more relevant when we consider the articulation of cultural imaginaries whereby female stars become models of identification and recognition for their female audiences, not merely a male fetishist object of desire. It is, however, through a strategy of failure that we can read the cinematic experience, where the word *failure* is not intended in a negative sense but comes to define the process of a desiring feminine embodiment that interacts with societal and cultural forces which also hinder its fulfillment. In cinematic terms, here *failure* translates into the tensions between passivity and activity, objectivity and subjectivity, directing and acting, and packaging and identifying. In analyzing the films of these female stars, failure refers to the tension between the directorial narrative and cinematic frame and the feminine embodiment (the confluence of corporeality and desire) that resists the controlling impulses of the former two. These tensions do not describe dichotomies with one element simply opposing the other because both are interlinked forces, part of a continuum, within a process of mutual interaction. Here, then, failure comes to stand exactly for the very impossibility of division and separation between these two forces and translates cinematically through a desiring feminine embodiment.

Embodiment

Before looking at the films in detail, let us clarify some terminology. I would argue that the idea of failure, as a continuum of tensions, is implicit in Elisabeth Grosz' notion of embodiment and especially embodiment as alterity, the latter defining the continuum between external and internal forces and the negotiation between the body and its own difference. This is a conception of subject and subjectivity, where the "subject is no longer an entity" but "the effect of pure difference that constitutes all modes of materiality," and subjectivity is no longer "the combination of psychical depth and corporeal superficiality but a surface whose inscriptions and rotations in three-dimensional space produce all the effect of depth."[6]

Positing a non-binary discourse between body and subject, Grosz refers to "prosthesis," a concept that encompasses a materiality external to the body, such as clothes, make-up, spectacles, artificial limbs, or plastic surgery. These are agents of modification that affect the body and, therefore, the process of embodiment. In this sense, the notion of prosthesis complements the notion of alterity that comes to signify the capacity of the body to "open itself to prosthetic synthesis, to transform or re-write its environment, to continually augment its powers and capacities through the incorporation into the body's own space and modalities of objects."[7]

Fran Martin and Larissa Heinrich remind us there already exists an extensive scholarship on body studies in China.[8] Speaking of the body's place within a social hierarchy, Tani Barlow states that historically Chinese bodies have played an important part in the cultural formation of social representation,[9] as the enactment of social, historical political power relations that reflect intrinsic contradictory hierarchies.[10] In a recent book in Chinese, Zhang Caihong talks about the need to survey the last hundred years of Chinese cinema using an ideology of body politics, especially when it comes to exploring stardom.[11] Thus, even though Chinese concepts, such as *qi* (energy), *sheng* (body), and *feng* (wind) do not easily translate into English, I see a strong similarity between the Chinese conception as described by Barlow and Grosz's idea of embodiment as a continuum between external social-cultural forces and the internal construction of the subject and its identity.

If embodiment equates alterity with failure, connecting tensions embedded within the other and forming inseparable parts of a continuum of interacting tensions rather then releasing them, then desire becomes essential here as it activates this dynamic process of interaction. Merleau-Ponty's definition of desire, in the words of Galen A. Johnson, well explains how desire facilitates this "enactment of embodiment" through visibility:

> The figure of the fire and the notion of desire it indexes refer us to genesis and growth, the possibility for new and renewed expression of the visible *simpliciter*. The longing within the Flesh is the longing for vision, new and renewed comprehension and self-comprehension... The desire or *conatus* of the Flesh is the demand for expression, the demand that the world be brought forth over and over again into visibility.[12]

Merleau-Ponty's definition of desire focuses on the longing for vision and expression. Like Grosz, he defines desire is rooted in materiality—the physicality of the experiences in which individuals can only desire through their bodies and individuals are defined as body-subjects. An important aspect of his definition is the link between desire and the longing for visibility, for self-expression, which metaphorically could stand for the subject-body longing to find a place in existence. In a Chinese context, though, some might assume desire has less impact in a culture imbued with Confucian principles of obedience and hierarchical order. Highlighting changes occurring in the 1990s, however, Lisa Rofel identifies ways in which a contemporary, cosmopolitan Chinese society has nurtured ideas of desire and consumption:

> Consumption is about embodiment, embodying a new self. At the heart of embodiment is *desire*. A properly cosmopolitan self is supposed to be desirous and this desire is supposed to be open and constrained.[13]

It is interesting to note that she associates the changes in Chinese society with the interaction between embodiment and desire, whereby consumption instigated by desire is about new embodiment, the construction of a new self. In particular, her 1990s survey suggests that the New China is full of desiring subjects who have a *desire* for anything: sex, possessions, modernity, etc. I can see a parallel between the paradigm consumption-desire that is historically specific to the 1990s and still present in contemporary China and the process of early 20th century modernity—specifically the modernity of 1930s Mainland China and 1950s Hong Kong. In his definition of a sensorial history of cinema, in particular the cinema from the early of 20th century, Zhang points to a kind of cinema that prompted a "visceral response, not only an intellectual one".[14] In both 1930s China and 1950s/1960s Hong Kong, this kind of response was directly linked to a desire for change and for modernity, and it produced a kind of cinema based on a desire for new bodily experiences, ones which facilitated the process of a *feminine* embodied cinematic experience.

Embodiment, Desire, and Stardom—Alterity, Prosthetics, and the Face

In line with Merleau-Ponty's idea of desire as the "demand for expression, the demand that the world be brought forth over and over again into visibility,"[15] I demonstrate that Chinese female stars, embodying a desire for self-expression, evince a "will to power." However, the desire that facilitates the process of embodiment can become tantamount to failure because desire is the part of embodiment that drives and actualizes these tensions. In this case, we can draw out the tensions between the stars' own desire or those of the cinematic industry of their times. Failure occurs because neither of these forces can prevail and none of these tensions can be resolved.

This becomes clearer if we look at the idea that stardom implies both the life on screen and off screen. The prominent film scholar Richard Dyer describes stardom as a phenomenon that not only includes the stars' films but all that is around the films—promotions, public appearances, biographies, and other materials—because "star images are always extensive, multimedia, intertextual."[16] However, here, I focus on these female actresses, their bodies, and their performances on screen as agents through which desire and failure are expressed and explored.

I especially highlight three aspects of this dynamic enactment: alterity, deriving from Grosz' idea of continuum and the way these actresses connect themselves to others within the cinematic environment and texture of the filmic experience; clothing and prostheses, referring purely to the ways stars' bodies are adorned and modified by external material elements, like costuming, make-up, or any element purely connected to their superficial appearance on screen; *qi* energy, to examine how stars' faces and facial expression convey embodiment. The first two aspects are connected to the definitions outlined above while the third element is located in a Chinese cultural context. To develop the idea of *qi*, I want to refer to John Hay's description of the human body in ancient Chinese art as being "dispersed through metaphors that locate it in the natural world by transformational resonance and brushwork."[17] This process results in embodying the cosmic-human reality of *qi*, or energy.[18] In his definition, *qi energy*, the cosmic energy that is the "obvious embodiment of humanity"[19] is expressed through the face, which was the most predominant part of the human body in ancient Chinese art painting. In my analysis of these films, I draw an analogy among humanity, *qi*, and the face to stress the prominence of facial expression in cinematic terms as an agent in the process of enacting a desiring feminine embodiment, a process made possible through the function of close-ups.

In this regard, I am developing my argument after the early 20th century film theorist, Béla Balázs, who considered close-ups as "the poetry of the cinema"[20] because they emphasize dramatic moments, "convey a subjective image of the world,"[21] and grasp "the mood of human being"— the "expressive look of a human gaze."[22] Following his description, we can regard close-ups as the perfect moments for stars to intervene in directorial dominance. However, while Balázs attributes close-ups to the director's sensitivity, I would rather emphasize how actors can express their desire for visibility through the potential spaces afforded by close-ups. Moreover, it is important to notice that in the films in questions, we experience an excess of close-ups, an excess of attention given to the stars of these films that can be explained, especially in Lin Dai's case, by the fact that many movies from this period were created around the star figure.

Ruan Lingyu

Ruan Lingyu began her career at the prominent Ming Xing Studio and moved in 1930 to Lianhua studio, where she made her most memorable movies. Social and media pressure and personal issues led her to commit suicide by poison in 1935 at the age of 24. Her life story was made into a movie in 1992, Stanley Kwan's *Center Stage*, where she is played by Maggie Cheung. Here, let us consider two of these films—*The Goddess* (*Shennü*, 1934) and *New Women* (*Xin nüxing*, 1935). *The Goddess* depicts the story of the life of a sex worker in urban Shanghai who tries to raise her small son as a single mother. It is a tragic tale that portrays a woman in a changing modern Shanghai. As she tries to free herself from a low-life pimp, she accidentally kills him and is convicted of murder. Her son is adopted by the head teacher of the school, and she makes a solemn promise to leave her son's future in the hands of the respectable headmaster and never to contact him again. *New Women*, directed by left-leaning director Cai Chusheng, is inspired by the real life story of Chinese actress Ai Xia, who in the 1920s committed suicide, as did Ruan later, due to the media pressure of the time. It depicts Wei Ming, a teacher turned author, who loses her job at the school and is forced into sex work to provide for her daughter. Both films touch upon sex work and deploy the stories of single mothers struggling for their independence and well-being. While one can argue that *The Goddess* focuses on themes of embodiment, sex work, and victimization, one can also surmise how *New Women* draws out the relation between redemption and liberation through bodily annihilation.

The Goddess

The Goddess explores questions of alterity through its portrayal of the different personas the character embodies: her public life as a sex worker and her private life as a caring mother. These two personas are represented by how Ruan uses the body in movement and in connection to its own external environment, which, in the case of the public persona, is an imprint of the social order that imposes itself on the individual. The restraint of the social order is visible in the public persona through the actress's ability to show subtly that her walking and behaving as a sex worker is a performative act, a sign of her alterity. This is highlighted by her practiced expression of preoccupation that blocks her appearing seductive or sensual.

The clip of her walking in the street at the very start of the film shows the actress embodying neither confidence nor sensuality. The iconic moment that is also reproduced by Maggie Cheung in Kwan's biopic sees Ruan walking slowly towards a stool, sitting on the stool, and once finally seated, cautiously and firmly tampering a cigarette on her leg. This image of the battered, proud sex worker aware of her own vulnerability and trying to keep calm and in control of the situation is evocative and highly performative. In contrast to this public sex worker persona, Ruan expresses a naturalness and lightness in the scenes where she attends to her child. Even in these beautiful, charming episodes, though, Ruan performs the natural presentation of a preoccupied mood, a presentation her face continues to project throughout the film. In this regard, Zhang stresses how Ruan's technique was far more natural than that of rival actress Hudie, who beat Ruan for the title of movie queen in 1933, according to reviews of the time.[23]

The character wears a typical *qipao*, a long slim-fitted all-covering dress that is neither revealing nor provocative. While this sartorial device alters the actress's sensuality, it also, using Grosz's word, "solidifies" Ruan's body, codifying her slender figure, untouched and unscarred by hard labor, pregnancy, and even poverty. Ironically, her slender embodied beauty, solidified by her clothing, imprisons her. In this regard, it is interesting to note that the *qipao* did not establish itself until the 1930s[24] yet came to define the Chinese model of modern femininity and "became a stage for debates about sex, gender roles and aesthetics, the economy and the nation."[25] In fact, by evolving into shorter dresses and developing longer side splits, it challenged traditional puritanical Chinese aesthetics and was a sign of women's changing roles in contemporary society. However, in the case of this film, the cinematic embodied experience is neither liberatory nor emancipatory. This is similar to Wong Kar-wai's use of the *qipao* in his lyric film *In Mood for Love* (*Huayang nianhua*, 2000), which is set in 1960s Hong Kong. Peter Brunette and Linda Chapple have read in this film a sign of corporeal absence. Despite the highly sexual appeal of Maggie Cheung as main character Li-zheng, Brunette describes a sexuality that is celebrated but repressed,[26] while Chapple suggests the dress becomes a metonymic replacement for the absent female body.[27] In the absence or the repression of embodied female sexuality, we can see Ruan's embodiment. (As I show below, Chapple's reference to the fetish having replaced the female body by making it absent does not apply to *The Goddess*.[28]) In Ruan's case, unlike Wong Kar-wai's constant attention to the *qipao* and to the female body, the focus gradually shifts from the whole body to the face as the center of attention in this film. This is not to say that Ruan failed to embody what Hjort calls "consistently present sexual energy"[29]. It is her expressivity, combined with her incredible ability to change moods in a controlled and leveled manner that distinguishes her from other actresses of the time.[30] Clearly at the center of the film, Ruan advances a naturalistic style antithetical to the "pantomime tradition of exaggerated and clearly codified gestures" Hjort sees as typical of the period.[31] As Berry and Farquhar suggest, the close-ups of her face are the "cinematic trope" that best represents Ruan,[32] and are a very frequent representation throughout *The Goddess* and *New Women*. As much as viewers empathize with the character, Ruan's acting emphasizes movements and behaviors suggesting performative layers imposed from the outside.[33] While putting societal codes and constraints on display, she also projects a deeper and more profound inner conflict that always accompanies the character on the journey of the film. A similar kind of acting can be found in *New Women*.

New Women

Much as prosthesis and ornament accrue psychological and corporeal significance in *The Goddess*, the "sombre black *quipao*" of *New Women*, as Kristine Harris argues, turns the characters into a neutral silent image, a sort of "tabula rasa that will be appropriated with meaning only after it has been appropriated by others."[34] In particular, Harris observes how the director Cai makes use of a split subjectivity to tell Wei Ming's story. Highlighting alterity across times, the film's formal techniques connect Ruan's image playing Wei Ming in the present to her former self by crosscutting between contemporary settings and images of those settings in the past. This strategy is particularly noticeable in the driving scene that shows Wei Ming on her way to the cabaret. Yet, even under the weight of these directorial devices, Ruan expresses the character's turmoil with restraint and pathos throughout the film. These scenes are more emotionally charged than similar scenes in *The Goddess*.

Interesting differences emerge while comparing the facial expressions of the final scenes from the two movies. In *The Goddess* we see the character walking in her cell, already convicted. At first, she looks disheveled in the cell as she sits slowly on a bench. A close-up shows her slightly in profile, on the left of the screen, as she puts away her hair from her face. She slowly turns her head to the right and looks up. As she has often done in the film, she starts to smile. Looking into the distance, she smiles, evoking longing and hope. This is soon framed by the intertitle that follows it and defines her feeling: "She will imagine in serenity the future of her son." In *New Women*, among several intercuts with outside scenes, we see Wei Ming lying in a hospital bed after having attempted suicide, fighting for her life, refusing to die. Surtitles are added on the scene; they say that she wants to live on. After that, we see her peacefully giving in to death, but the surtitle is still imposed over her. The main difference is in the use of intertitles. In *The Goddess*, the first intertitle is imposed as an outside commentary on the last scene of the film, pronouncing a moral judgment on the character. In the second film, the surtitle (rather than an intertitle) expresses the character's inner voice and communicates a sense of injustice, almost as if the deathly situation has been imposed upon her.

However, as in *Goddess*, in between several crosscuts, between hospital clips and outside clips, the actress offers a pronounced smile of serenity. This moment after a scene of pain and turmoil, where the character had expressed her desire to survive, makes her death and her serenity rather artificial and untruthful. Such artificiality also runs through the structure of the film. In particular, we experience the didactic tone of the use of crosscuts between Wei Ming's dying image and the outside situation, a classroom full of women portrayed as militant activists and described as instigators for change and freedom, suggesting she dies in response to their activism. In this instance, Harris sees "Wei Ming's subject position, a despairing 'individual new woman' being eclipsed by the collective 'we' of utopian new women,"[35] and argues that this eclipsing crosscut attempts to highlights the political content of the film, which seeks to condemn the media and the society at large for the character's death.[36]

Looking beyond the narrative elements of the films to the facial expressions of the actress, we can see the embodied display of desire, the actress's attempt to transcend her own punishment in *The Goddess* and her own mortality in *New Women*. In this regard, Laikwan Pang asserts,

> The director seemed so anxious to portray Wei Ming's image and thoughts in order to express himself, to the extent that his own control threatened to vanish. Instead, the gaze of the spectator was returned directly with the gaze of the character, forcing the spectator to recognise his/her position hitherto defined by the cinematic apparatus.[37]

Pang describes the character's gaze in a clip with no surtitle, looking back to spectators to make them aware of her embodied pain and desire for transcendence. Her look signifies her plea to make her situation visible. This contrasts with the victimhood imposed on her by the director through the film's narrative structure. In this regard, we should note that the portrayal of the conditions affecting women in movies like *Goddess* and *New Women* was part of the "nation-building project of 1930s"[38] that presented at least the pretense of emphasizing female emancipation and equality. Defining this directorial approach and the attendant ideological agenda as part of a larger nationalistic project, Chris Berry and Mary Farquhar refer to prominent Chinese-American scholar Rey Chow's argument that female figures stood in for China, a nation that was feminized by critics before and during the Sino-Japanese war (1937–49). During that period, China was characterized as weak and fragile and in need of being rescued.[39] Activists and intellectuals on the Left used a concern over the material conditions of women to promote change and re-establish a greater and stronger China.[40]

However, more clearly in *The Goddess* and less so in *New Women*, the failure provoked through the tension among the narrative, the director's gaze, and the actress's expressiveness emphasizes spectators' failures to resolve the tension between the competing ideological agendas in play. The longing and desire for visibility expressed through the two final close-ups of both films, as the actress attempts to make the characters' condition as long-suffering woman visible, remain longing and desire. *New Women* is less subtle, and failure is less visible because the narrative seems to prevail as the filmmaker uses the character to express his own feelings and ideology.

Harris refers to critical responses connecting Ruan's character Wei Ming to Ibsen's character Nora in *A Doll's House* (1879). Some comparisons see the impossibility of either Nora or Wei Ming fulfilling women's independence after leaving their marriages.[41] Such critical response and Ruan's own death can be seen as extensions of this impossibility. However, in her performances on screen, Ruan redeems herself: her images and what is left from the filmic imaginary signify a failure that is made visible through a "will to power," a desire for visibility. Even though this desire intersects with (and does not overturn) the power of directorial discourse, it succeeds in reacting against what Dai Jinhua describes as the male gaze's necessary objectification and subjugation of the female body.

Linda Lin Dai

With Linda Lin Dai, we enter a different Golden Age, that of the Mandarin musical in Hong Kong. After 1949, the "colossal war"[42] between the big two studios, Shaw Brothers[43] and MP & GI, prompted "the center of…cinematic gravity" to move from Shanghai to Taiwan and Hong Kong.[44] Many within the industry began to label Hong Kong the "new 'Eastern Hollywood,'"[45] where the main players were not male stars but female stars.[46] Two of Lin Dai's films from this era, *Cinderella and Her Little Angels* (*Yushang yanhou*, 1959) and *Love Without End* (*Buliao qing*, 1961) are especially interesting for our considerations of failure and masquerade.

Cinderella portrays Wang Danying, an orphan/dressmaker turned model who helps set up a fashion show, organized by a local boutique and used as a fundraising event to save her orphanage from closure. While the others orphan girls continue sewing women's dresses for the local boutique, Wan Danying becomes the star of the fashion show. The film also focuses on one of the local boutique clerks, Lin Fu (Chen Hou) and his obsession with a mannequin that resembles Wang Danying. *Love Without End*, with its iconic film score, is a love story between an immigrant girl who makes her way as a singer in the Hong Kong nightclub scene and the son of a rich entrepreneur about to go bankrupt. This is a tragic film that emphasizes the embodied aspects of sacrifice, illness, and death.

Lin Dai became a beauty symbol of the time. Her career started after a producer spotted a picture of her displayed at a hairdresser. However, the image she presented remained dynamic, rather than simply replicating that first appearance. In fact, her looks became increasingly Westernized as she adopted thick round eyebrows that made her eyes look bigger and an even thicker and rounder hairstyle, attributes that earned her the title of "Elisabeth Taylor of the East." At the time, mainstream Hong Kong cinema was preoccupied with representing a cosmopolitan modern reality marked by images of women "fit to travel," who "travel to fit," as Brian Hu, adopting the language of sociologist Jennie Molz, has suggested.[47] These two expressions denote how the feminine body was at the center of the film industry and was used to embody a cosmopolitan model that reflected contemporary cultural trends in a modern and multicultural city[48] that wanted to project a cosmopolitan image of its local film industry as it sought foreign capital and collaboration.[49] Most importantly, the cinematic production of this period promoted a "utopic image of the mobile Chinese woman fit to travel and travelling to fit in global cosmopolitan culture"[50] in its attempt to reach transnational Chinese communities situated in South-east Asia and beyond.

Hu's article on cosmopolitanism raises questions about Lin Dai's stardom as a vehicle to build a series of cosmopolitan films and cosmopolitan networks.[51] Therefore, we could argue that there are some strong similarities with Ruan's situation: both actresses' personas were part of bigger schemes—ideological and political in Ruan's case and commercial in Lin Dai's. One way or the other, the two actresses were used by their respective film industries. I would argue that, as in Ruan's case, Lin Dai's performances on screen create tension among these industrial, transnational concerns, the narrative cinematic discourse, and a desiring feminine embodiment. This unresolved tension marks a paradigm of failure, whereby Lin Dai is not a mere mannequin in the hands of the film industry but, like Ruan, an agent who embodies a "will to power" made visible through these two films.

Cinderella

In the case of *Cinderella*, an MP & GI production, masquerade and performativity are elaborated through the trope of the doll that defines a society in transition between modernity and tradition. The orphanage and the boutique are built around a strict hierarchy, embodied in the figures of the owner of the boutique and the headmistress of the orphanage. However, Lin Dai's character, Wang Danying, becomes a disruptive agent, disobeying the headmistress' order not to take part in the fashion catwalks, even though she does so only to help raise money for the orphanage. However, as we see from the cinematic images that present a joyful and excited Lin Dai (the actress) in the musical spectacle of the fashion shows, Wang Danying (the character) seems to find pleasure and fulfillment in these activities and becomes the center of adoration and attention. The sense of freedom and the filmic spectacle of the catwalks contrast with the austerity and the poverty of the orphanage. It is a contrast between the old and the new, between restriction, sacrifice, and responsibility and freedom, amusement, and colorful cosmopolitanism.

Danying is at the core of the narrative and of the filmic experience from the start of the movie. Fascinatingly, she is made the sign of her own alterity as her relationship with her own otherness is made material and embodied by the presence of the life-sized, wooden mannequin that accompanies her throughout the film. This relationship with alterity is especially prevalent in three sequences. The first sequence, in the early stages of the film, portrays the dream of Li Fu, one of the boutique clerks. In the dream, the wooden mannequin turns into a live woman, played by Lin Dai, who wears a large white dress reminiscent of a fairy or princess costume. In this collective dream, the other clerks and the other orphans also appear. Medium-wide shots are used to portray a big stage and focus on a long staircase in the middle, which is used by the characters for a musical dance routine. In the second sequence, as the clerk seems to interact with the mannequin, he does not notice Danying, who enters and starts playing tricks on him. This is a playful sequence, with some fast crosscuts, a cat and mouse situation where Danying follows the clerk around the shop. The third sequence is highly comical and bizarre. The orphanage headmistress searches the boutique for Danying and as she sees the mannequin that looks like her standing in the middle of the shop, she confuses the mannequin for the real Danying, who is also posing as a mannequin. Danying stands so still and both the mannequin and she look so much alike that the headmistress fails to recognize the real woman.

Throughout these three sequences, the image of the wooden body mirrors the image of the fleshly body, connecting aspects of mimesis, corporeality, and desire. Lin Dai plays her character and sometimes plays the mannequin, multiplying the reproduction of the images of her character and the reproduction of her own image. On screen, then, reflection and desire appear to project in multiple directions, and the character, the mannequin, and the actress all appear to take pleasure in their displayed embodiment through their projected mimesis. Playing at being other to oneself, rather than being authentic to oneself, seems more pleasurable here.

But what does the young clerk's dream imagine about desire? At the start, we learn he prefers the doll-woman to the real woman. However, in the second sequence, the actual Danying reveals herself to him, but the clerk needs a long time to recognize her. The same can be said about the headmistress and the clerks from the boutique. Desire and a fascination with mimicry delay them from discovering reality, hinder their recognition. They are not simply confused but fail to recognize Danying for so long because they are awed by her reproduction. Furthermore, this fascination with replication evokes a similar pleasure from the spectator as Lin Dai enacts this carnal desire with regard to alterity. Here, alterity is defined by a kind of desire materialized through a constructed display of the body. It is not so much about authentic visibility as it is about mimesis in the form of masquerade.

This relation between mimesis and masquerade, then, highlights the dynamics of prosthesis. In fact, it can be suggested that Lin Dai's embodies the act of mimesis and that the apparent pleasure she experiences is mirrored again in her experiences of the fashion sequences. In the case of the fashion scenes, we can identify not only the spectacle of display but also the embodiment of different personas and different cultures through the various clothing and catwalk routines. On the catwalks, Lin Dai juxtaposes wearing traditional clothes from different countries with donning trendy contemporary fashions and through this sartorial display embodies cosmopolitanism and modernity.

The first fashion show presents Lin Dai alone on the catwalk, where she wears more modern than traditional dresses, demonstrating a neutral modernity. Lin Dai displays herself as a sex symbol who wears trendy and sensual outfits. The second is a wedding fashion show presenting pan-Asian dresses. This time, the clerks from the boutique are also on the catwalk taking part in the show, and the number of people on stage combined with shots of the diegetic audience becomes a display of family and communal traditions. Neither fashion show is a traditional catwalk, but both become nearly developed dance routines. In the second show, characters' movements are slightly more restrained and controlled because they wear more traditional dresses. The scene explores a modern 1950s/1960s Hong Kong through fashion, where new social codes are still rooted in traditional codes. In early scenes set in the orphanage, before the fashion shows commence, the clerk engages in sartorial confusion, when he wears a modern dress while asking the women in the orphanage to reproduce a new design. Since this dress does not fit him properly, the women conclude that the dress cannot be made. The clerk's cross-dressing scene is an uneasy attempt within the film to navigate between an interest in the novelty of modernity and an attachment to more conventional sartorial codes. This unease is also highlighted in the contrast between the chaste representation of the female orphans and the display of Danying's outgoing personality. In fact, it is in Draying's case that fashion as prosthetic device has deep social and cultural significance because fashion has the potential to change Danying from a naive orphan into a knowing sex symbol. Again, fashion stands here for the embodiment of a cosmopolitan modernity that has affected Danying the character and Lin Dai the actress. Unlike the *qipao* in Ruan's films, clothing does not imprison the character/actress but liberates her. Furthermore, the final show of Chinese wedding dresses and traditions portrays Lin Dai as outside of tradition. As the characters prepare for the final catwalk show, Lin Dai hints that the head ornament is too heavy for her. The significance is plain here: traditional marriage will not fit Danying the character and Lin Dai the actress. As with the clerk, some clothes just do not fit. He could not wear a certain dress, but neither can she. Through sartorial devices, the scene suggests Chinese traditions do not fit modern, cosmopolitan people and their lifestyles, just as cosmopolitan urbanities do not confirm tradition.

If prosthesis and ornament expose cultural embodiment, they also can be limited to the level of spectacle and display through the act of masquerade. The latter is mainly present in the second fashion show, created by the intercutting between images of the show and images of audiences' fascinated reactions to the show. The intercutting highlights that these catwalks are performative displays and embody a sense of being-looked at from an external point of view. As Hu suggests, we need to note that the fashion show serves diegetic and non-diegetic functions.[52] In this sense, the film indeed makes palpable the irreconcilable tensions among modernity, tradition, embodiment through prosthesis, and masquerade.

The image of the doll is a material incarnation of Lin Dai's smile and becomes almost emblematic of Lin Dai as a movie star. The mannequin does not change but solidifies her Westernized traits and smile. Desire here is embedded within a societal structure by a collective desire that is well expressed through a sequence twenty minutes into the film. This sequence shows the orphans involved in preparing and sewing clothes in the orphanage. Danying walks from the boutique to the orphanage. The sequence begins with the camera moving from Lin Dai's feet to her head and then cuts to a close-up of her face. This image is intercut with a shot of the girls at their sewing machines, followed by a medium-shot of the headmistress looking at the girls. The intercutting continues until Danying arrives at the orphanage and starts singing with them. There are long sections within this sequence where Danying remains an individual thanks to medium and close shots, but her individuality is visually connected to the orchestrated choreography, accompanied by the non-diegetic music and the characters' singing. While this sequence connects the collective and the individual, through the display of the energetic playfulness of both, the figure of the headmistress stands diachronically opposed to such a resolution. As an individual figure standing against Danying, her image functions as regulator and main observer. If there is joyful desire for modernity and freedom in this film, it remains highly regulated.

In another sequence before Danying leaves for the first fashion show, we see her closing the gate of the orphanage. Combined with her melancholic singing, her act of closing the gate evokes a complex experience of liberation and frustration. As she closes the gate, she does not go out but locks herself in. This act is related to the strict rules imposed by the headmistress, and the growing tension between the two characters unfolds throughout the narrative. It is important to notice that this sequence rejects close-ups in favor of medium and wide shots.

In comparison with Ruan's films, the desiring feminine embodiment appears more strictly regulated and framed within the narrative structure of *Cinderella*, but only because the tensions between modernity and tradition, the global and the local, and desire and restraint are stronger and more identifiable than in Ruan's case. Yet, I would argue that this difference does not make Lin Dai merely the silent other, the image of a naked woman rushing through the dreams of men, as suggested by Dai Jinhua. The orchestrated display of the fashion shows and the wooden doll (as the clerk's fetishized object of desire) does not merely objectify Lin Dai and her character. Against such gestures toward objectification, these signs of female spectacle negotiate the relations among embodiment, mimesis, and desire, as the mimesis of Lin Dai in the doll sequence and the close-ups of the actress's facial expressions embody a different kind of desire—a desire for visibility projected through the gaps available in the central role she plays in the film's narratives. Furthermore, her star status is not merely the function of a manipulative, male-dominated film industry, for her stardom, like that of Lin Dai, largely stemmed from her huge popularity amongst female spectators.

In this sense, the clerk's desire for the mannequin can be read as a fantasy and as such is no longer available to him in the diegetic reality of the film. The clerk remains a passive character throughout the film and the objectifying desire he projects toward Lin Dai remains unfulfilled. The woman that plays the game of mimesis is a woman with her own ability to disguise herself and play tricks on the people around her. Therefore, she is not merely a doll; she is the center of the narrative, of the cinematic focus, and is literally the star within both the diegetic and the non-diegetic cinematic experience.

Love Without End
A tragic story of love and death, *Love Without End* presents Li Qingqing, a Chinese immigrant turned nightclub singer, who sacrifices her own reputation for love, only to fall victim to a fatal illness. She saves her lover from bankruptcy by becoming an escort to a rich villain and decides to stop seeing him. These motifs and the end to the film echo *The Goddess* and to the sacrifice the mother makes to see her son grow up under better circumstances, while the sudden death of the character recalls the death in *New Women*.

Masquerade is part of the narrative apparatus that is kept together by the musical element of the movie, the diegetic singing of Qingqing performing on stage, and the occasional non-diegetic off-stage singing and musical score. As in *Cinderella*, masquerade creates a distance, a layer of artificiality between the tragic events of the film and its own characters; however, the film's over-dramatic tone affects spectators to evoke an embodied emotional response to the events depicted. Again, as the film traces the tension between two contradictory elements, a sense of failure emerges between distance and proximity.

Here, the focus is not on costume but on the environment—the Hong Kong nightclub setting—where Li Qingqing (embodying the tensions running through the film) changes from an immigrant to a nightclub star. As in *Cinderella*, her appearance, sartorial adornments, and new hairstyle add to the displayed changes in her status and personality. However, the film more clearly connects Li Qingqing's changes to the external environment by crosscutting between her body and the Hong Kong nightclub scene where she is located. The Hong Kong night scene is depicted as trendy, modern, and above all cosmopolitan with little tension between the old and the new. Lin Dai's face shows even more prominent Western traits with a fuller hairstyle and more pronouncedly rounded, big eyes. This Westernization is mirrored by her status as wealthy star: her apartment with a maid in a high-spec decor is a display of independence and sophistication. At a narrative level, failure and victimization are present in the shallow masquerade of wealth and sophistication that contrasts with her unhappiness. She is coerced into sex work to help her lover, and she ultimately suffers from poor health and dies at the end of the film. Material plenty, then, fails to bring pleasure in the film.

In terms of alterity, her connection with the outside, the film focuses on the performative musical display that sometimes hints at the psychological pathos portrayed through Lin Dai's graceful and emotional acting. The emotional chore of the character is often masked by its own alterity, the overdramatic pathos, typical of musical dramas, in the character's relationship with the cinematic space. In one of the sequences, after she breaks with her lover, she readily expresses her sadness at seeing the man she loves attend the show and enjoy the company of other women. The relation of the character with the space creates an added layer that highlights how Lin Dai's acting is artificially overdramatic. This artificiality is unlike Ruan's subtle emotional display but I suspect does not reflect Lin Dai's acting skills; rather, the performative excess presents a challenge to the narrative codes of a genre that regulates and circumscribes Lin Dai's character with a tightly framed *mise-en-scène* that reflects her rigidly framed emotional range. Beyond this, by looking at close-ups of her face, we can recognize two different kinds of expressions: one of longing and one of despair. The scene at the window when she says goodbye to her lover and sees him walk away in the distance develops a commentary on alterity by juxtaposing these two expressions with regard to pathos.

In the rooftop scene that begins the film, close-ups focus on her pensive yet cautiously optimistic mood as she waits for her uncle to arrive at the nightclub. This scene comments on the longing for a future where tears evoke sadness coupled with hope. As in the case of Ruan, there is a sense of heaviness that signals the end of the film in a subtle manner. The end of the film portrays Qingqing's waiting for death: as she sits alone on a cliff by the seaside, we see a medium shot against the backdrop of a seashore. The roiling sea mirrors her internal turmoil. At the end, she has the same pensive look she had in the opening scene; however, the use of a medium shots and the intercutting of to the images of the turbulent waters add a dramatic element to the scene, signifying the pain and loneliness that the director imposes upon his narrative—a narrative Lin Dai does not fully embody. Instead, she embodies a similar if not more subdued sense of longing, the same sense of desire she expressed during the rooftop scene.

If we draw a comparison with the last scene in *New Women*[53] we can see in both characters the same desire to transcend death, to express their own "will to power." However, as we see from the last scene, unlike *New Women*, the actress is not the last person we see on screen. As her lover arrives at the house on the cliff, she is already dead and we no longer see her; the last shot is that of a tumultuous sea. Within the tensions between directorial scheme and embodied performance, this might mean that after making her face larger than life, the director decided to obliterate her image and eliminate her ability to articulate her desire. In this sense, the failure is blocked because the narrative structure seems to prevail over the actress' articulation of desire. Without tension, there is no failure. In this sense, in the end, Qingqing would remain another character subject to Dai Jinhua's model of femininity that has been objectified, fetishized, and silenced by the cinematic gaze.

Nonetheless, if we read Lin Dai's performance in the context of her career as an actress, especially in comparison with earlier films, even *Cinderella*, we might alter our opinion. Lin Dai's established screen image was that of happiness, health, and optimism.[54] With one of her first films, *Singing Under The Moon (Cui Cui*, 1953), Hong Kong warmed very easily to her representation of the naïve, long-braided fishing girl image,[55] and since then, until her later, darker films, this image of happy and cheerful performer, judged by some to display an aggressive sexuality,[56] had stayed with her. It is with this intensely dramatic role in this film, though, that Lin Dai managed to step out of the star image that her spectatorship, the industry, and she had created. This renegotiation of her image through the film increased her fame and brought her broader recognition as a film star. Therefore, as it was the case for *Cinderella*, despite the tensions among objectification, corporeality, and desire in the film, Lin Dai's star persona remains an image of empowerment and model of strong feminine self-fashioning at the center of the film industry's attention at that time. This model of tense embodiment, then, is a far cry from the silenced woman suggested by Dai Jinhua's analysis.

Concluding Remarks

These films demonstrate a paradigm of failure through the presence of contradictory forces, mainly among the narrative discourse, the director's vision, the ideological constraints of the society, and the actresses' desire for self-expression, their "will to power" displayed through their embodiment—their movements, gestures, and above all, facial expressions. As we have seen in Ruan's case, not even death, on-screen or off-screen, was able to block the actress from projecting this strong desire for visibility. Almost thirty years later, Lin Dai projected a vision of cosmopolitan modernity that remains unresolved at the level of the tensions between masquerade and performance in the musical genre. Lin Dai had more difficulty sustaining these tensions. Still, through her deployment of a strategic image, iconic off-screen persona, and position in the center of the cinematic experience, she was able to project her corporal desire for visibility. One cannot say for sure whether the differences between the two stars and the increasing difficulty for self-expression are due to the changes in the film industry, which had become increasingly dominated by a capitalist urge to reach larger audiences, almost on a global scale. However, the legacy of these two stars that surpassed that of the people who had controlled and manipulated their on-and off-screen personas remains a testament to their desire not to be silenced and objectified into the the feminine Other.

Notes

1 Dai Jinhua, ""Ren, Gui, Qing": yige nurende kunjing" in Dai Jinhua ed. *Wuzhong fengjing: Zhongguo dianying wenhua, 1978-1998* (Beijing: Beijing da xue chu ban she, 2006), 146.
2 Tani E. Barlow, *The Question of Women in Chinese Feminism*, (Durham: Duke University Press, 2004), 336.
3 One of the exceptions is expressed in a recent Chinese book on body politics and female stardom in which scholar Zhang Caihong affirms that female stars can break through male hegemony, even this is not thoroughly explained. "Films are glorious day dreams, and movie stars on the screen, embody a collective desire for this glorious dream. One could argue that by showing the materiality of the body while displaying its beauty, they [female stars] are also exploring opportunities to break male suppression and gain some ground against male hegemony, ideology and consumerism to find the possibility to construct their own femininity." Caihong Zhang, *Shenti Zhengzhi:bainian zhongguo dianying nü mingxing yanjiu* [Body Politics: a study of one hundred years Chinese female stardom] (Beijing: Zhongguo guangbo dianshichubanshe, 2011), 12.
4 Zhang Zhen *An Amorous History of the Silver Screen: Shanghai Cinema, 1896-1937* (Chicago: University of Chicago Press, 2005), 37.
5 Ibid., xvi.
6 E.A. Grosz, *Volatile Bodies: Toward a Corporeal Feminism* (Bloomington: Indiana University Press, 1994), 208.
7 Ibid., 188.
8 F. and Heinrich Martin, L., "Introduction to Part 1. Thresholds of Modernity" in *Embodied Modernities: Corporeality, Representation, and Chinese Cultures*, ed. F. and Heinrich Martin, L. (Hawaii: University of Hawai'i Press, 2006), 5.
9 . A. Zito and T.E. Barlow, "Body, Subject, and Power in China," in *Body, Subject, and Power in China*, ed. A. Zito and T.E. Barlow (Chicago: University of Chicago Press, 1994). 10.
10 Ibid.
11 Zhang, *Shenti Zhengzhi:bainian zhongguo dianying nü mingxing yanjiu* 9. (It is interesting to note that he uses ideas from Merleau-Ponty, Dyer among others but his approach is mainly historic and socio-political.)

12 Galen A. Johnson, "Ontology and Painting 'Eye and Mind," in *The Merleau-Ponty Aesthetics Reader: Philosophy and Painting*, ed. M. Merleau-Ponty, G.A. Johnson, and M.B. Smith (Evanston: Northwestern University Press, 1993), 54.
13 L. Rofel, *Desiring China: Experiments in Neoliberalism, Sexuality, and Public Culture* (Durham: Duke University Press, 2007), 118.
14 Zhang Zhen *An Amorous History*. 35.
15 Johnson, "Ontology and Painting 'Eye and Mind." 51.
16 Richard Dyer, *Stars* (London: BFI Pub, 1998), 3.
17 John Hay, "The Body Invisible in Chinese Art? ," in *Body, Subject, and Power in China*, ed. A. Zito and T.E. Barlow (Chicago: University of Chicago Press, 1994), 44.
18 Ibid.
19 Ibid., 52.
20 Béla Balázs. *Béla Balázs: Early Film Theory: Visible Man and The Spirit of Film*. Ed. Erica Carter. Trans. Rodney Livingstone. (New York, NY: Berghahn, 2010), 41.
21 Ibid., 44.
22 Ibid., 41.
23 Zhang, *Shenti Zhengzhi:bainian zhongguo dianying nü mingxing yanjiu* , 91.
24 A. Finnane, *Changing Clothes in China: Fashion, History, Nation* (New York: Columbia University Press, 2013), 149.
25 Ibid., 141.
26 P. Brunette, *Wong Kar-wai* (Urbana, IL: University of Illinois Press, 2005), 91.
27 Linda Chapple, "Memory, Nostalgia and the Feminine: In the Mood for Love and Those Qipaos," in *Millennial Cinema: Memory in Global Film*, ed. Amresh Sinha and T. McSweeney (New York: Columbia University Press, 2011), 217.
28 Ibid.
29 Mette Hjort, "Ruan Lingyu: Reflections on an Individual Performance Style," in *Chinese Film Stars*, ed. Mary ed Farquhar, introd, and Yingjin Zhang, *Routledge Contemporary China Series (Routledge Contemporary China Series): 51* (London, England: Routledge, 2010), 39.
30 Yan Dai, *Meili yu aichou- yige zhenshide Ruan Lingyu* [Beauty and sadness - the real Ruan Lingyu] (Beijing: Dongfang chubanshe, 2005), 189.
31 Ibid., 37.
32 Chris Berry and Mary Farquhar, *China on Screen: Cinema and Nation*, Film and Culture (FiCu) (New York, NY: Columbia UP, 2006), 121.
33 Mette Hjort, "Ruan Lingyu: Reflections on an Individual Performance Style," 38.

34 Kristine Harris, "The Goddess: Fallen Woman of Shanghai," in *Chinese Films in Focus II*, ed. Chris Berry (Basingstoke, England: Palgrave Macmillan, 2008), 281.
35 ———, "The *New Woman* incident - Cinema, scandal, and spectacle in 1935 Shanghai " in *Transnational Chinese Cinemas: Identity, Nationhood, Gender* (Hawaii: University of Hawaii Press, 1997), 286.
36 Ibid.
37 Laikwan Pang, *Building a New China in Cinema: The Chinese Left-Wing Cinema Movement, 1932-1937* (Lahnam: Rowman & Littlefield Publishers, 2002), 128.
38 ———, "The making of a National Cinema: Shanghai Films of the 1930s," in *The Chinese Cinema Book*, ed. Song Hwee Lim, et al. (Basingstoke, England: Palgrave Macmillan, 2011), 61.
39 Berry and Farquhar, *China on Screen: Cinema and Nation*. 121.
40 Lianhua, which produced *Goddess*, employed Leftist writers and directors (ibid. 119). Cai Chusheng was a Leftist director.
41 Harris, "The *New Woman* incident--Cinema, scandal, and spectacle in 1935 Shanghai ". 288.
42 Ibid., 25.
43 According to Poshek Fu, the Shaw Brothers studio was "the most successful in constructing a pan-Chinese film culture in the diaspora by means of its little-challenged domination over the Chinese-language cinema industries around the world" Poshek Fu and Lane J. Harris, *China Forever: The Shaw Brothers and Diasporic Cinema*, Popular Culture and Politics in Asia Pacific (Popular Culture and Politics in Asia Pacific) (Urbana, IL: U of Illinois P, 2008), 8.
44 J. Yang and A. Black, *Once Upon a Time in China: A Guide to Hong Kong, Chinese, and Taiwanese Cinema* (Atria Books, 2003), 24.
45 Ibid., 25.
46 Y. Zhang, *Chinese National Cinema* (London & New York: Taylor & Francis, 2004). 177.
47 Brian Hu, "Star Discourse and the Cosmopolitan Chinese: Linda Lin Dai Takes on the World," *Journal of Chinese Cinemas* 4, no. 3 (2010), 190.
48 K. Louie, "Introduction," in *Hong Kong Culture: Word and Image*, ed. K. Louie (Hong Kong: Hong Kong University Press, 2010), 7.
49 Hu, "Star Discourse and the Cosmopolitan Chinese: Linda Lin Dai Takes on the World." 190.
50 Ibid., 204.
51 Ibid., 190.
52 Ibid., 192.

53 This is no coincidence. Scholar Sam Ho talks about the echo of a "hard-edged Shanghai of the past" in *Love Without End*. L. Kar, F. Bren, and S. Ho, *Hong Kong Cinema: A Cross-cultural View* (Lanham: Scarecrow Press, 2004), 234.
54 Ibid.
55 Liu Shu, "Hongyan boming di meili jiaren——Lin Dai (Unlucky Marie--Lin Dai)," *Dazhong dianying* (2004), 50.
56 Emilie Yueh-Yu Yeh and Darrell William Davis, *Taiwan Film Directors: A Treasure Island*, Film and Culture (FiCu) (New York, NY: Columbia UP, 2005), 234.

CHAPTER 4

Trans-mothering on Singapore's Siniticate Screens
Jun Zubillaga-Pow

Abstract

In this chapter, the figure of the Chinese "trans-mother"—a cross-dressing male actor in the role of a mother—is used to delineate how the fictitious role of the mother/widow has remained shackled to the familial function of providing parental guidance and infantile protection. In particular, I argue that recent Singapore cinema has retained the patriarchal tradition of portraying mothers as homely as they have been in most Shakespearian and Asian theatre. How these values of vice and virtue come to be embodied in the cross-dressing actor is no more an anti-feminist oddity within Siniticate screen culture than it is in everyday reality. To the extent that these narratives have also shaped cultural perceptions of trans-mothers and their role in bringing up children vis-à-vis family values, cinematic representations have little altered dramatic, filmic, or social attitudes. I contend that how Singaporeans have come to accept or tolerate gender-changing and cross-dressing as a staple of contemporary visual media is analogous to the constant negotiation of linguistic semantics in the global nation-state, where ethnicities, languages, and beliefs have been disentangled and made homogeneous.[1]

Keywords: Mothering, Singapore, Siniticate, Togetherness-in-Difference, Trans

那你几时会跟妈咪 穿一样的衣服？还有，以后你要我叫你爹地还是妈咪？"

(When will you start dressing up like Mummy? And from now on, should I call you *Daddy* or *Mummy*?)

In a recent film, 世界末日 (*Judgment Day*), the father character tells his family that he wishes to undergo gender reassignment surgery after the alleged doomsday.[2] Immediately, the social and cultural conservatism of Singapore's society become unbearable for him. Even his teenage son reacts angrily by castigating him on his preferred form of parental address. From this single articulation, we can observe readily the gendered transgressions between normative behavioral and linguistic differences in the colloquial interchange between spoken Mandarin and English. In this instance, the politics of naming bodies and behaviors is predicated on the semantic effects of multilingual reception and can produce a unifying and conformist discourse for the viewers. The English subtitles on the film or television screen also subject the audiences who do not understand spoken Mandarin to a specific Chinese cultural milieu, or what I will refer to as a "Siniticate," that creates a mono-semantic understanding of the narrative. Taken as an integral cinematic experience, what we are seeing is, in one way, a brief episode from the everyday life of a Singaporean family conversing in a hybridized Sinophonic tongue and, in another way, fictitious life stories being received visually and aurally by a Siniticate community of people who are being affected by the acts and speech-acts on screen.

Togetherness-in-Difference: The Chinese, the Sinophone, and the Siniticate

Singapore is made up of a multiethnic and cosmopolitan populace; its residents are equipped with different language skills and infused with a diversity of cultural influences. In contrast to the way most authors write about Singapore, my preference here is to frame the residents in the multifold of the Chinese, the Sinophone, and the Siniticate. Critically, I aim to deconstruct the Sinocentric positioning of Singapore as a single discursive entity that is based on a lingual or ethnic divide. Although it is of minority practice or subaltern norm to frame Singapore as a Sinophone state, the potpourri cultures of native-born Eurasians, Chinese, Indians, and Malays have hitherto not been considered vis-à-vis reception studies. To set the scene of this argument, I provide a critique of the work of two scholars in literary and political ontology. Shu-mei Shih has written on the inherently "polyphonic and multilingual" nature of Sinophone articulations in Asia and the Pacific. Observing the lingual experiences of people who have grown up or are living in Singapore qua "a settler society,"³ she asserts,

> The fact that postcolonial Singapore adopted English as the national language, even though each citizen is allowed free expression in his or her "mother tongue," appears to have led to Sinophone Singaporean literature's not being the most desired language of expression. Sinophone literature in Singapore therefore must always be examined in conjunction with Anglophone literature in the context of the official policy of multlingualism [sic] on the one hand and the globalizing impulses of the city-state on the other.⁴

As a multilingual Singaporean myself, I have had different experiences with languages and literatures; I have encountered situations and circumstances with more acute sensitivities. Likewise, it is not uncommon for the recent cohort of Singapore-based writers to include non-Sinitic languages—such as English, Malay, Tamil, and colloquial euphemisms—in Sinophone Singapore literature. Whether this displacement or replacement of the Sinophone on Chinese tongues is an expression of desires or otherwise, the call for plurality and openness has been gaining ground, most notably in the past decade, with the rise of an audible milieu of bilingual actors and singers.⁵ However, I am less convinced by Shih's reasons for Singapore's Sinophone literature to be examined alongside Anglophone literature only. As much as the top-down imperative has been for citizens to embrace a multiracial and multilingual culture, the interchange of languages and rituals has been lackluster.⁶ When such a phenomenon is applied to local film reception, the racial and cultural compartmentalization becomes much less pronounced than the Barthesian binary of genotype and phenotype for two reasons.

First, residents who view Sinophone screen material are not necessarily fluent in the language themselves. Drawing on Marshall Hodgson's differentiation between the Islamic and the Islamicate, I will appropriate a similar methodology between the Sinitic and the Siniticate, where the second category includes people who are influenced by the acts and customs of those categorized under the first term.⁷ For example, the inscription of a Chinese figure by an ethnic-Malay poet in the *pantun* could be considered as a Siniticate act, as would a non-Muslim eating *halal* food be participating in the Islamicate culture.⁸ Let us look at an example from one of the early trans-mothering television shows, 女人40当自强(*The Power of Women at 40*).⁹

In one scene, the Chinese trans-mother explains to her *peranakan* guest how her curry is different from the *peranakan* curry due to the potpourri of different regional flavors. She calls it a *chaperanakan* and *humperanakan* curry. The first portmanteau is the combination of the Hokkien word *cha* meaning "to mix" with the Malay word *peranakan* meaning "locally born," while the second neologism is derived from the Cantonese word *humpahlang* meaning "everything" and *peranakan*. With this semantic hybridism, we get an uncanny sense of a locally invented curry that includes everything that is edible. In such a Siniticate world, different Sinitic and non-Sinitic languages are intermixed to create localized meanings. Hence, we see even in a city with a Sinitic majority, there remain a sizeable number of non-Mandarin-speaking compatriots who access, enjoy, and influence Sinophone material. Indeed, here in Singapore, non-speakers have as much influence as others who understand the language, regardless of our varying abilities to speak it.

Second, the political scientist Kenneth Paul Tan analyzes the work of Singaporean filmmakers Jack Neo, Eric Khoo, and Royston Tan as constituting a Marcusian "mono-dimensional" cultural genotype. However, I contend that the economic and aesthetic returns of their output are heterogeneously distributed because there remains a correlation between how the filmmakers pitch their films toward a target audience and how the reception of these textual phenotypes, regardless of the filmmakers' political orientation, is already calibrated by the erstwhile authoritarian policies.[10] For example, Eric Khoo's productions appear to be geared toward winning accolades at international film festivals, while those by Jack Neo are composed and localized specifically with the suburban "heartlanders" in mind. Unlike Frederic Jameson's and Aijaz Ahmad's interrogations of Third World national allegory or Françoise Lionnet's and Shih's creolization of theory and literature, Singapore's product differentiation cannot be forced into a simple dichotomy between the national and the minor or between the cosmopolitan and the popular.[11] To perceive Singaporean articulations as such is to undervalue the intellectual acumen and socio-cultural mobility of the devoted producers and patrons of the arts.

Against such conceptualizations, a better frame for studying the reception of film, literature, and other cultural artifacts from Singapore is Ien Ang's ontological concept, "togetherness-in-difference."[12] A media scholar, Ang coined the phrase to describe certain instances of human migration and acculturation. She suggests that the socio-cultural complications that arise as people get more and more entangled with each other can be classified as a living form of "togetherness-in-difference."[13] Such an optimistic stance is helpful for the purpose of apprehending the different corporeal and verbal hermeneutics involved, especially in times of explicit projected or internalized transphobia and linguistic xenophobia. This conceptualization is better for two reasons. First, Shih and her disciples have already assured us that linguistic communities are open and open to changes. Second, trans scholars have stipulated subjects "in trans" to be "composed of multiple determinants... that allows for their reassembly."[14] Both imagined communities are constructed on ontological differences and similarities that allow their members to embody "togetherness-in-difference." Whether as gendered or linguistic communities, membership of the respective group is always already heterogeneous and hybridized, and their constitution ensures their organization remain fluid and dynamic. Both Shih and Ang have beseeched us to move against and beyond diasporic thinking, and trans academics and activists have advocated a similar position pushing gender and sexual discourses beyond binary heteronormativity.

Therefore, I see the coherence in applying the principles of "togetherness-in-difference" in thinking about the screen as a unifying resource and trans-mothering as a unifying event. Screens can perform "togetherness-in-difference" gradually and eventually with advanced development in digital technology. Reaching beyond the perimeters of film and television screens, screen studies scholars would be greater informed by expanding their field to include the smaller and larger digital display forms such as the mobile phone, the computer screen, and the video wall. The diversification of screen resources would provide a better understanding of how mediated culture is further distributed and received. Media production and regulation in Singapore is a case in point, and extensive scholarship has been and continues to be carried out.[15]

Correspondingly, a critical survey of trans-mothering presented here is possible if and only if we subscribe to the philosophy of "togetherness-in-difference." It is this "togetherness-in-difference" that enables the conjunction of trans feminism to be represented visibly and audibly by male actors playing mothers. Their screen presences are mediated by file-sharing websites such as YouTube on the small screens and by public advertisements and political messages broadcast on the large screens found in shopping malls or on the public transport network. In the next section, I will discuss the vicissitudes of trans-mothering from stage to screen in Singapore as a globalized discursive center vis-à-vis theatrical and cinematic history.

Trans Portrayal on Singapore Screens: Liang Ximei, Karen Neo, and Aunty Lucy

Scholarship focusing on transgender topics in Singapore over the past twenty-five years remains rare. One of the earliest psychological surveys was recorded by the psychiatrist Dr. Tsoi Wing Foo, who has been providing medical services to transsexuals since 1973.[16] The only other scholar who has researched transgender issues is Ho Chi Sam, whose master's thesis questions the politics of representing transgenderism in recent Singapore press and "LGBT communities."[17] Contingent to the Singapore context, I use the term "trans" throughout this chapter to refer to people who identify with neither one nor the other gender.[18] To be sure, the visibility of such gender pluralities has brought about the discipline as well as discrimination against trans people in Singapore since the nineteenth century. According to pre-Independence oral history records, there were already men of *peranakan*, or mixed Chinese-and-Malay descent, dressed in fashionable women's costumes. These were a large group of men who were "gentle and friendly" and would "make lovely flowers." They would "laugh and giggle like women" and "walk *lenggang lenggang*"—a form of sashaying.[19] In addition, historical sources also confirm the presence of female impersonators in local theater shows from the late nineteenth century. These shows, known as *wayang peranakan*, were dramatic enactments of everyday life, and the performers were descendants of the Chinese who settled in Malacca, Singapore, and other parts of the Malay Peninsula during the seventeenth century. They were strongly influenced by the Malay opera, the *bangsawan*, which itself was adopted from Persian theater passing through Mughal India.[20]

Following the routes taken by the Islamic missionaries, I have tried to trace the genealogy of trans-mothering from Iran to the Malay world, but to no success. Despite the presence of daughters, mothers, and wives in the *Ta'zieh* plays dating from the nineteenth century, we do not know anything about the professional status of women or male persons taking up female roles within Persianate theater history. My conjecture concerning Islamic practices is that male-to-female cross-dressing would be more *haram* than having female actors on stage, who would otherwise be given weeping or non-speaking parts.[21] To be sure, this position excludes the secular transvestite theater of East Java. Nevertheless, the *peranakans*, who did not convert to Islam, have less of a cultural barrier in cross-dressing on stage or in everyday life. Eventually, the international clout of feminism provided the impetus for a female protagonist to take center-stage in the form of Stella Kon's

1983 play, *Emily of Emerald Hill*. However, film scholar Kenneth Chan has accused the drag interpretation by one of Singapore's *peranakan* actors, Ivan Heng, of being an anti-feminist act of "political self-erasure."[22] By repositioning a male body on stage in the role of a matriarch, Heng crosses gender but also crosses out gender as his appearances in 2001 can be considered a return to Singapore's Euro-American or Asian-derived patriarchal theater practices. It is no surprise that cross-dressing on Singapore screens eventually proliferated with the end of the Cold War and the start of state neoliberalism. Working within a Siniticate "melting pot," local filmmakers have been widely influenced by theater and cinema that involves trans characters, including sixteenth-century English renaissance theater, seventeenth-century Venetian opera and Chinese Kunqu, eighteenth-century Japanese Kabuki and Javanese Ludruk, British comedy skits, and Hollywood road movies.

One of the most multitalented Singaporean actors is Jack Neo (梁智强), who has acted in a handful of cross-dressing roles on large and small screens since the early 1990s. While his portrayal of Grandma Neo (梁婆婆) has been well-received, it is his interpretation of Liang Ximei (梁细妹), a housewife and mother of two sons, in a television series called 女人40当自强 ("The Power of Women at 40") that has resonated most with the everyday lives of working class television viewers. Aired during the 8 p.m. primetime slot, this specific creation by Neo is the accumulation of more than a decade of making comedy on Singapore's only Mandarin-language television channel. The popularity of the comedy sketches could be attributed to Neo's grasp of the mundane aspects of motherhood. In the episode mentioned earlier, the trans-mother prepares a bowl of chicken curry for her *peranakan* neighbor. When the trans-mother goes to fetch her guest, her sons collude with their friend to steal some of the food. To compensate for the amount they have eaten, her sons mix paper napkins into the curry and pretend that the curry actually contains *kohlay chai*, or "cauliflower" in the Hokkien dialect. The sons distract the guest from putting the curried napkins into her mouth by making barking noises and knocking over a coat rack. As a traditional act of parenting, the trans-mother reprimands her children in front of the guest. From this scene, a social hierarchy is drawn between mother and children; the children should behave themselves when there is a visitor or when adults are discussing important matters. Otherwise, the trans-mother would "lose face" or be ashamed for failing to maintain her authority over her children and within the household. Such instances of judging the efficacy of a trans-mother's parenting skills based on her children's conduct are not without historical and trans-cultural precedent. We can compare this scene to a similar one in *Little Britain*, a British sketch comedy show created by David Williams and Matt Lucas. In an episode from 2003, the trans-mother, Sandra Patterson, brings her son Ralph to a casting audition. She boasts to another parent about how much effort she has to put in to bringing up her son so that he can become a famous actor. She then shows her bare feet to indicate the sacrifices she has been willing to make by how willing she was to sell her shoes in order to send her son to a private school. Both examples—of the characters of Liang Ximei and Sandra Patterson—suggest how the trans-mother, even more than the mother, must prove herself as exceptionally wise, coming up with witty ideas to promote or protect her children, and exceptionally self-sacrificing, surrendering all she has as the children grow up. At the end of the scene with Liang Ximei, when the truth of the curry becomes clear, the trans-mother complains to the audience how making peace with her neighbor and taking care of her sons at the same time is such a brain-racking problem: "哎哟, 真的是伤脑精哦!" (Oh! It's really brain racking!)

In 2011, the Singaporean trans-mother appeared in the format of a transnational road movie. In 笑着回家 (*Homecoming*), Jack Neo portrays the widow Karen Neo, who goes bargaining for an eighty-eight dollar *yusheng*, or raw fish salad, before accompanying her son on a coach from Singapore to Kuala Lumpur for their Chinese New Year reunion dinner. Along the way, they meet various stock challenges like the coach breaking down, the son overdosing on sleeping pills, and Neo using a winning lottery ticket to pay for a young woman's flight home. The actions and reactions of the trans-mother in the film often serve as role modeling for her son (as well as for the audience) in the sense that her struggling through adversities and ultimate success in arriving home preach the virtues of perseverance and altruism. Uncannily, the role of Karen Neo resembles that of the American character Bree Osbourne in *Transamerica* (2005). Osbourne is a pre-op trans person who goes on the road after bailing her son out of jail. At one point in the film, the trans-mother lectures her son: "Eat your vegetables, you might want to use a fork, just an idea," "Fasten your seatbelts… You don't have to say 'like' [in the middle of a sentence]," and "please don't put your feet up on the dashboard." She embodies kind-heartedness and empathy when she takes on a shamanic hitchhiker, who later steals her car and hormone pills, and an aboriginal rancher, whose romantic agency comes to emphasize the independence and asexuality of the trans-mother.[23] In the personalities of Liang Ximei, Karen Neo, and Bree Osbourne, we witness the selfsame representation of a strong and righteous woman, always ready to sacrifice her personal pursuits for the welfare of her child or children.

Another distinct personality trait of the trans-mother on screen is that of transferring her own wishes onto her children. Generative wish fulfillment is a major topic in psychoanalytic discourse whereby the child's desire has been theorized as the desire of and for the mother. For example, Sandra Patterson brags in *Little Britain* that her son's talent has been inherited from her. Being too old to be an actor, she says she is now going to live out her dreams through her son. On learning that her son has not been selected, Sandra's jealousy and self-consolation become more comedic than sympathetic, though, because the trans-mother is portrayed as merely relying on her child to fulfill her own unaccomplished, "unrealistic" ambitions. The imposition of a mother's own desires onto her child has obvious theatrical precedence, such as the trans-mothering performed by Constance in Shakespeare's *King John*. All in all, these empirical and philosophical examples support Nancy Chodorow's claim that the act of mothering becomes a form of reproducing socio-political and cultural ideologies, including those of more general parenting practices, which I will return to in the next section.

In a similar vein, the film *Aunty Lucy*也灌篮 (2009) deploys a parental role model in which the trans-mother lectures her son to be as magnanimous as she is. Aunty Lucy is portrayed as a stereotypical homemaker who is shamelessly parsimonious—as indicated by her catchphrase "So embarrassing!" As a cost-conscious widow raising two teenage children, she collects vouchers or coupons and patronizes vendors giving out free samples. When one of her neighbors falls ill and is left with an expensive medical bill, Aunty Lucy forms a basketball team with her "aunty" friends to win the ten thousand dollar prize to help their unfortunate neighbor. Later, she realizes that she will be playing against her son and a childhood nemesis. However, Aunty Lucy knows that she must win the competition and the prize money to save her neighbor's life. In the end, she disregards her son's sporting ambition and uses the situation instead to teach her children about the importance of altruism and self-sacrifice.

Such a spirit of selflessness is also portrayed in another episode of Jack Neo's comedic series when Liang Ximei teaches her son the proverb, "救人一命，胜造七级浮屠" ("saving someone's life is better than building a seven-story pagoda"). However, this time the lesson in altruism is deployed more for comic effect than to teach ethics as the trans-mother mispronounces "浮屠" (*fú tú*; pagoda) as "糊涂" (*hú tú*; muddle), creating an uncanny, ironic joke. Because her version of the proverb states, "saving someone's life is better than being extremely muddleheaded," the trans-mother thinks that she needs to save someone's life in order to prove her mental clarity and stability.

The semantic mistake refers equally to the son's state of confusion and immaturity, which the trans-mother is trying to alleviate. Again, the comedy of the trans-mother circulates within an economy of mis-recognition and mis-articulation.

When the trans-mother is not in any comic situations, her keenness to ensure good social behavior is emphasized. A case in point is how the trans-mother, compared to the mother, is ever more determined to teach her children not to repeat her mistakes.[24] For example, Lucy was very mischievous when she was younger and would hang out with her two delinquent girlfriends. In order to deter her children from following in her bad ways, she now monitors her children's online activities and encourages her nerdy daughter to be more feminine and friendly with her male classmates. Such generational contrasts in sexual expression allow viewers to reflect on the different constructions of femininity embodied by Aunty Lucy and her daughter. Instead of bullying innocent boys, Aunty Lucy's daughter dresses up to attract the boys' attention. Granted, her daughter's sartorial choices could be strategic on several levels. Because she needs to distract the rival basketball players from winning the game, she decides to dress provocatively. This particular scene takes on the meaning of how every trans-mother's wish is realized by her own daughter, and for Aunty Lucy's daughter this transference is accomplished through acquiring her trans-mother's celebrity qualities, by being attractive, popular, and famous. In this regard, the trans-mother's childrearing efforts can be perceived as reflecting the social mores of the milieu and not on radical trans-gender performativity. Simply by teaching her daughter how to be feminine, the trans-mother maintains patriarchal gender roles and also upholds the beacon of being a good and wise mother ("贤妻良母"), where the Confucian diktat disguises the maternal self-sacrifice as altruism with the paternalistic superego ruling over her subjected ego.

State Fatherhood Redux

We know from Nancy Chodorow's landmark writings that the imperatives of the patriarchal have continued to perpetuate the role of women as submissive wives and mothers to their husbands and children, especially when the nuclear family unit becomes modeled on the capitalist reproduction of labor.[25] Although the definitions of masculinity and femininity have changed significantly since Chodorow discussed them in the late 1970s, to a large measure the traditionalism and domesticity of the maternal has continued to bear the burden of the parental, including the majority of childrearing responsibilities. At the heart of Chodorow's critique is her assertion that "families create children gendered, heterosexual, and ready to marry," and I would not hesitate to extend her claim to the expectations of reproducing the very culture and ideology inculcated in them.[26] There are also the second wave feminists who neither accept the subjective state imposed by the father nor reject the mental and physical totality mothers bear for choosing motherhood.[27] This subversion is most pronounced in how Rebecca Walker has framed her own decision in choosing maternity: "There would be loss and mourning involved in the abandonment of my preoccupation with myself, but so far, even with the unbelievable pain and complications, it is all gain."[28] The spatial-temporal and affective experiences of motherhood return to interrogate feminism.

In fact, the very embodied elements of motherhood that convey the energy and gratification of kinship between mother and child have been most suited to the dramatic arts and have been represented on stage and screen. However, due to the social and political organization of theatrical traditions, female roles taken up by men have given birth to the figure of the trans-mother. Albeit played by male actors, the renewed presences of trans-mothers across times and spaces have become significant arbitrators of contemporary maternal feminist scholarship. Often, the trans-mother is portrayed as someone who is decisively in control of relations, morally self-righteous, and naturally the matriarch of the family—whether her spouse is present or not. In her own way, she expresses her care and concern for her children, but her gestures and actions remain always conservative and paradoxically self-centered, despite her avowed altruism. On stage or on screen, the trans-mother is always represented in the act of mothering. She is always first and foremost a mother, portrayed as having to make an extra effort to impress upon audiences her maternal potential in simultaneously keeping the house and family in order and raising the children. She must always prove she can pass as a mother, and the conduct and future of her children are almost always determined by her capabilities. Yet, she is always shown in flux and ultimately is appraised by an ideal set of mothering skills and responsibilities. Thus, her mothering is always misrecognized in the end.

To paraphrase the art of mothering, the cultural spectacles of our times have continued to be myopic of the maternal, and a critical "reconciliation" of mothering and art on the stage and screen continues to be urgently overdue.[29] This particular stance is also very much aligned with the contemporary aspiration for the feminist third wave, where the trans-mother on stages and screens, in contrast to the biologically-sexed mother, is more committed to maintain the powerful bond with her child through trials and tribulations, beyond life and death.[30] Unlike the catatonic and hysterical mothers from numerous plays and films, such as Sam Mendes's *American Beauty*, the figure of the trans-mother is self-critical and does not hesitate to acknowledge her shortcomings. The trans-mother defies the post-maternalist call for a reduction of motherhood responsibilities by reasserting the patriarchal norms of mothering and adhering to a neoliberal form of matrophobia, or the reified fear of becoming one's mother.[31] Essentially, the male actor is not afraid of becoming the trans-mother situated between her own father and son. The trans-mother is dramatically different from, say, Irina in Chekhov's *The Seagull*, who dares to stand up to her unfaithful lover and suicidal son and asserts her unruliness. Instead, the trans-mother is unabashed to repent if and when she has failed to meet the ideal against which she is judged, which means she is always confronted by such a cycle of failure and repentance.

Akin to the representations of homosexuality in French queer cinema from the late 1990s, cross-dressing mothers on the television programs or in the theaters of Singapore continue to display the comedic or malefic personas given them.[32] Roles perceived as virtuous in everyday societies are seldom assigned to male-to-female trans actors. The figure of the mother, who is one such character rarely seen on stage and screen, is often portrayed in caricature, emphasizing the misrecognition, desolation, and abjection of her nature. This is why the Confucian expectation of maternal duties, known traditionally as "贤妻良母" (good wife and wise mother), stands alienated from the real and imaginary lives of trans-women both on and off screen.[33] Whatever screen time is allocated to the trans-mother, she is most often portrayed as a widow (or a single parent for all intents and purposes) who ultimately fails as a caregiver and ultimately fails in her quest to nurture a strong mother-child relation. Indeed, it is possible to suggest that, for a ubiquitous genealogy of trans-mothering to become plausible, these trans-mothers on screens must invest in Ien Ang's principle of "togetherness-in-difference." Regardless of their different mothering and linguistic capabilities, these trans-mothers can gain a contingent agency only by realizing their positions and being together through togetherness.

The cultural force of this "togetherness" is what makes the pervasive imposition of "state fatherhood" upon the private lives of citizens relevant to this thesis.[34] With the continual decline in Singapore's total fertility rate since the 1970s, the government's involvement in directing both public and commercial media content remains pronounced to the discerning eye. Following Geraldine Heng and Janadas Devan, a patriarchal representation of mothers on television and cinematic screens has persisted from the early years of industrialization to the current neo-liberalization of the city-state. With the implementation of the classification system for films and videos by the Media Development Authority in 1991, the above-mentioned films and videos have received varying degrees of censorship.[35] In addition to the narratological characteristics discussed here, the marked differences in classification statuses pose a political rejoinder to the proclamation made by Heng and Devan that Singaporean women are "by definition always and already anti-national."[36] On the contrary, the performances of Liang Ximei, Karen Neo, and Aunty Lucy become a didactic message from the state to local women citizens that motherhood and mothering is for the greater good of the consumerist society and the precarious nation. The reproduction of mothering is not restrained to the embodied relations between mothers and children but, in fact, disseminates state-sanctioned values and ideologies beyond the maternal kinship through the socio-cultural multipliers of the screens and media networks they permeate.

From the perspective of cultural history, the reappearance of the trans-mother from early modern transnational stages on early twenty-first-century screens can therefore be considered as a revival of patriarchal systems, Confucian or otherwise. Not discounting the fact that the directors and scriptwriters of these films and television programs are all men, I can imagine that certain actions deployed on screen and certain affects evoked through these experiences are as real as those experienced by actual transgender mothers within and without the Siniticate world.[37] What remains definitive about the list of trans-mothers, cinematic or otherwise, is their willingness to stand together in differences and audiences' willingness to interpret them in such a manner. Regardless of class, language, nation, or race, these trans-mothers have been conscripted to perform the maternal as assigned to them by the economic and socio-political hegemonies of different times. As exemplified through the theory and practice of trans-mothering across time and space, the trans-mother, as opposed to the mother, is always and already nationalized and undermines transnational feminist justification. For Jack Neo and Dennis Chew, the irony of becoming local star celebrities as well as role models for everyday mothers on the street is the obliteration of the trans-mothers' dynamic personhood and personality. That said, my hope is for these trans-mothers to embrace a definition of motherhood and maternity that is not determined by any single entity or authority but by their diverse localities and embodiments.

Notes

1. I would like to thank the editors, Brian, Mary and Wai Siam, as well as the audiences at the Global History and Culture Center, Warwick University and the International Gender Studies Center, Oxford University, for their constructive comments on earlier versions of this chapter. Throughout this chapter, I am using "trans" rather than "trans*" based on a historical trajectory and not contemporary readings.
2. 世界末日 (*Judgment Day*), directed by Ong Kuo Sin (王国燊), released on April 18, 2013, rated PG-13 with "Some Mature Content."
3. Shu-mei Shih, Introduction to *Visuality and Identity: Sinophone Articulations across the Pacific* (Berkeley: University of California Press, 2007), 28.
4. Shu-mei Shih, Introduction to *Sinophone Studies: A Critical Reader*, ed. Shu-mei Shih, Brian Bernards, and Chien-hsin Tsai (New York: Columbia University Press, 2013), 13.
5. E. K. Tan, "Transcending Multiracialism: Kuo Pao Kun's Multilingual Play *Mama Looking for Her Cat* and the Concept of *Open Culture*," in *Sinophone Studies: A Critical Reader*, ed. Shu-mei Shih, Brian Bernards, and Chien-hsin Tsai, 315–25 (New York: Columbia University Press, 2013).
6. See the press interview with Syed Farid Alatas in "Singapore Is Not Yet Truly Multicultural," *The Straits Times*, November 9, 2011, A14.
7. Marshall G. S. Hodgson, *The Venture of Islam: Conscience and History in a World Civilization* (Chicago: University of Chicago Press, 1974).
8. For example, one of the early Malay *pantun* goes: *Anak Cina menjual kesuma, / Mari dijual di Pekan Rabu; / Makan tak lalu tidur tak lena, / Teringat tuan saban waktu.* [The Chinese girl sells flowers, / She sells them at the Wednesday market; / I eat without appetite and sleep without rest, / Constantly thinking of my beloved.]
9. *The Power of Women at 40* (女人40当自强), directed by Jack Neo, Television Corporation of Singapore Channel 8, 1997–2003.
10. Kenneth Paul Tan, *Cinema and Television in Singapore: Resistance in One Dimension* (Leiden, Netherlands: Brill, 2008).

11 Aijaz Ahmad, *In Theory: Classes, Nations, Literatures* (London: Verso, 1994); Françoise Lionnet and Shu-mei Shih, *The Creolization of Theory* (Durham, NC: Duke University Press, 2011).

12 Ien Ang, "Together-in-Difference: Beyond Diaspora, into Hybridity," *Asian Studies Review* 27, no. 2 (2003): 141–54.

13 Ien Ang, *On Not Speaking Chinese: Living between Asia and the West* (New York: Routledge, 2001), 200.

14 Shu-mei Shih, "Against Diaspora: The Sinophone as Places of Cultural Production," in *Global Chinese Literature*, ed. Jing Tsu and David Der-wei Wang (Leiden, Netherlands: Brill, 2010), 45; Quoted in Helen Hok-Sze Leung, "Trans on Screen," in *Transgender China*, ed. Howard Chiang (London: Palgrave Macmillan, 2012), 184–85.

15 For examples, see Kokkeong Wong, *Media and Culture in Singapore: A Theory of Controlled Commodification* (Cresskill, NJ: Hampton Press, 2001); and Terence Lee, *The Media, Cultural Control and the Government in Singapore* (New York: Routledge, 2010). The Media Development Authority and the Infocomm Development Authority are two of several public statutory boards that control and monitor the flow of information within the nation-state via economic and political management. Whether perceived as propaganda or pornography, licensing and classification of cultural objects have inevitably become the practice de rigueur in Singapore's neoliberal culture industry. The efficacy of "togetherness-in-difference" can no doubt be turned on its head as a panopticon for public surveillance.

16 Tsoi Wing Foo, "The Prevalence of Transsexualism in Singapore," *Acta Psychiatrica Scandinavica* 78 (1988): 501–4; Tsoi Wing Foo, "Developmental Profile of 200 Male and 100 Female Transsexuals in Singapore," *Archives of Sexual Behavior* 19, no. 6 (1990): 595–605.

17 Ho Chi Sam, "Transgender Representations" (master's thesis, Department of Communications and New Media, National University of Singapore, 2010).

18 I follow the connotations of "trans" and "transing" as established by Susan Stryker, Paisley Currah, and Lisa Jean Moore in their introduction to *WSQ: Women's Studies Quarterly* 36, nos. 3 & 4 (2008).

19 Interview with Betty Seow and Guat Beng (Accession 1048/17), Oral History Center, National Archives Singapore.

20 On the origins of the *bangsawan*, see Tan Sooi Beng, *Bangsawan: A Social and Stylistic History of Popular Malay Opera* (Oxford: Oxford University Press, 1993).

21 Male-to-female cross-dressing in the Islamicate world is seldom present, except in sacred or folk practices. For example, see James Peacock, *Rites of Modernization: Symbolic and Social Aspects of Indonesian Proletarian Drama* (Chicago: University of Chicago Press, 1987), or Carl J. Hefner, "Ludruk Folk Theater of East Java: Toward a Theory of Symbolic Action," (PhD diss., University of Hawaii, 1994). Female-to-male cross-dressing, on the contrary, has existed since Medieval Islamicate literature. See Sahar Amer, "Cross-Dressing and Female Same-Sex Marriage in Medieval French and Arabic Literatures," in *Islamicate Sexualities: Translations across Temporal Geographies of Desire*, ed. Kathryn Babayan and Afsaneh Najmabadi (Cambridge, MA: Harvard University Press, 2008), 72–113. See also Rebecca Ansary Pettys, "The Ta'zieh: Ritual Enactment of Persian Renewal," *Theatre Journal* 33, no. 3 (1981): 341–54, esp. 353; Sasan Fatemi, "Music, Festivity, and Gender in Iran from the Qajar to the Early Pahlavi Period," *Iranian Studies* 3, no. 3 (2005): 399–416.

22 Chan, Kenneth, "Cross-Dress for Success: Performing Ivan Heng and Chowee Leow's 'An Occasional Orchid' and Stella Kon's 'Emily of Emerald Hill' on the Singapore Stage," in "Where in the World Is Transnational Feminism," special issue, *Tulsa Studies in Women's Literature* 23, no. 1 (2004): 43.

23 For other readings of *Transamerica*, see Gina Marchetti, "*Transamerica*: Queer Cinema in the Middle of the Road," in *Hit the Road, Jack: Essays on the Culture of the American Road*, ed. Gordon E. Slethaug and Stacilee Ford (Quebec: McGill-Queen's University Press, 2012), 198–213; and Kylo-Patrick R. Hart, *Queer Males in Contemporary Cinema: Becoming Visible* (Lanham, MD: Scarecrow Press, 2013), esp. 122–23.

24 See Kathleen Rowe Karlyn, *Unruly Girls, Unrepentant Mothers: Redefining Feminism on Screen* (Austin: University of Texas Press, 2011).

25 Nancy Chodorow, *The Reproduction of Mothering: Psychoanalysis and the Sociology of Gender* (Los Angeles: University of California Press, 1978).

26 Ibid., 199.

27 Julia Kristeva has been more forthcoming with suggestions on plausible ways of the maternal with her aim to traverse this socio-political "fantasy" by letting the rhythmic body of the feminine and the maternal dictate the meaning of her feminist

orientations between the instant and the eternal; see "Stabat Mater" in *Tales of Love*, trans. Leon S. Roudiez (New York: Columbia University Press, 1997), 234–63.

28 Rebecca Walker, *Baby Love: Choosing Motherhood after a Lifetime of Ambivalence* (New York: Penguin Books, 2007), 202.

29 Buller, Rachel Epp, ed., *Reconciling Art and Mothering* (Aldershot, UK: Ashgate, 2012).

30 Karlyn, *Unruly Girls, Unrepentant Mothers*, 227 and 232. Karlyn's example is the once ambivalent relation between Rebecca Walker and her mother, Alice Walker.

31 Margaret Morganroth Gullett, "Postmaternity as a Revolutionary Feminist Concept," *Feminist Studies* 2, no. 3 (2002): 553–72; Adrienne Rich, *Of Woman Born: Motherhood as Experience and Institution* (New York: W. W. Norton, 1976).

32 For examples, see Cristina Johnston, "Representations of Homosexuality in 1990s Mainstream French Cinema," *Studies in French Cinema* 2, no. 1 (2002): 23–31; and Kate Ince, "Queering the Family? Fantasy and the Performance of Sexuality and Gay Relations in French Cinema, 1995–2000," *Studies in French Cinema* 2, no. 2 (2002): 90–97.

33 Jennifer Robinson has delineated this socio-familial imperative in her essay "Dying To Tell: Lesbian Practice and Love Suicide in Early Twentieth Century Japan," in *Queer Diasporas*, ed. Cindy Patton and Benigno Sanchez-Eppler (Durham, NC: Duke University Press, 2000), 38–70.

34 Geraldine Heng and Janadas Devan, "State Fatherhood: The Politics of Nationalism, Sexuality, and Race in Singapore," in *Nationalisms and Sexualities*, ed. Andrew Parker, Mary Russo, Doris Sommer, and Patricia Yaegar (New York: Routledge, 1992), 343–64.

35 While *Homecoming* and *Aunty Lucy Slam Dunk* have each been given a PG (Parental Guidance) rating, *Little Britain* and *Transamerica* are rated M18 ("Suitable for persons aged 18 and above") and R21 ("Restricted to persons aged 21 and above") respectively.

36 Heng and Devan, 356.

37 In addition to Wong Kim Hoh's article, "When Papa Became Mama" (*The Straits Times*, September 8, 2008), a six-page feature report on trans-mothers is available on a local Chinese tabloid, 优1周加料, No. 180, May 30, 2005, 2–7.

CHAPTER 5

Coming Out in the Mirror: Rethinking Corporeality and Auteur Theory with Regard to the Films of Tsai Ming-liang
Hee Wai-Siam

Abstract

This chapter examines the intersection of auteur theory and corporeality in Tsai Ming-liang's oeuvre in the context of the development of the mirror experience. In doing so, it asks whether Tsai's films create a closet for Chinese queers or construct a queer performativity belonging exclusively to Chinese people. This article combines readings of the films with interviews with Xiao Kang and Tsai Ming-liang conducted by both the author and the media in order to rethink their mirror relationship from the perspectives of both men. Tsai borrows auteur theory to legitimize the process of creating Xiao Kang's queer body while simultaneously using the creation of the "Xiao Kang household" in the films to find an emotional home for himself as a permanent member of the Malaysian diaspora in Taiwan. Tsai's self-expression through his films has always been limited by his mirror relationship with Xiao Kang. This relationship forms a closet that refuses entry to language and understanding, and so narrative logic can only be minimized. The majority of audiences must rely on Tsai's extra-diegetic descriptions for their understanding of his films. The chapter concludes that Tsai's words, actions, and works have, much of the time, generated a closet for Chinese queers through the mirror experience. It is also a type of queer performativity belonging exclusively to Chinese queers.

Keywords: Corporeality, Coming Out, Queer, Auteur Theory, Tsai Ming-liang, Lee Kang-sheng

Corporeality: An Alliance with Auteur Theory

Tsai Ming-liang's films represent the deconstruction of the Chinese heterosexual family. They ask whether Chinese queers can come out of the closet, though if they overtly broach the subject of coming out at all, they do so only through the "coming out" story of Lee Kang-sheng (Xiao Kang 小康) from the body of a boy to the body of a man. Xiao Kang as an actor is the exclusive property of Tsai's films. He does not speak much. What remains of the character is his body and repetitive actions performed day after day: drinking, eating, wandering, defecating, masturbating, lovemaking, and sleeping. This is the repeated representation of Xiao Kang's corporeality in Tsai's films, an embodiment that redefines what we might mean by "coming out" in transnational Chinese cinema.

Even though more than twenty years separates *Rebels of the Neon God* and *Stray Dogs*, Tsai's films continue to use the original cast of actors, as Xiao Kang has grown from the young rebel of yesteryear to the miserable middle-aged man of today. The close relationship between Tsai and Xiao Kang kept people guessing right up until Tsai announced at the Venice International Film Festival that Xiao Kang was his "life partner."[1] Their relationship has provided fertile ground for the imagination of viewers of Tsai's work. As one critic has said of the on-screen Xiao Kang, "His erotic identity is complex: he desires and is desired by both men and women. His ambiguous and fluid sexuality is best understood as 'queer'…neither straightforwardly heterosexual nor homosexual. He has established a global following by developing a professional commitment to an individual auteur…."[2] Such reviews have led Tsai's viewers to ask after the sexual identity of Xiao Kang in real life.

Tsai has disclosed that Xiao Kang is not a *tongzhi*.[3] Xiao Kang has also clarified in an interview that he is not a *tongzhi* and expressed that he resists playing a *tongzhi* in Tsai's films.[4] These claims do not stop audiences from projecting their homosexual fantasies onto Xiao Kang on or off screen. When they watch Tsai's films, these audiences, like Tsai, seem to return to Lacan's "mirror stage" and project their desires onto Xiao Kang's bodily image. This "mirror experience" is the process of Tsai forming an alliance between corporeality and auteur theory. Does this constitute a Chinese queer closet to replace Anglo-American coming out politics or a queer performativity belonging exclusively to a Chinese cinematic experience? These questions seem unresolved, as much of Tsai's audience do not applaud Tsai's queer performativity.

Tsai Ming-liang's conceptual grounding flows from French auteur theory. He uses his films to mold Xiao Kang's corporeality into a "personal signature" of his own "self-expression." In recent years, the Taiwanese academic Peng Hsiao-yen has criticized Tsai's practices as "self-worship" that leads to endless "self-repetition."[5] Peng criticizes Tsai because he has "turned actors into moving props"[6] and emphasizes the aesthetics of the image over the other elements in his films. Critics such as Peng have criticized such priorities as an abandonment of "narrative logic"[7] and asked, "are self-expression and narrative logic necessarily exclusive to each other?"[8] Taking Peng's perceptive interrogation into account, this chapter examines "self-expression" in the light of the relation between desire and corporeality in Tsai's works. As Judith Butler points out, "The professionalization of gayness requires certain performance and production of a 'self' which is the constituted effect of a discourse that nevertheless claims to 'represent' that self as a prior truth."[9] This leads to several questions. In the process of creating his oeuvre, what relationships are created between his "self-expression" and gayness? What creates the conflict between "self-expression" and "narrative logic"? Have Tsai's films abandoned narrative logic? Finally, how do these questions link to Chinese queers "coming out"?

Toward the beginning of his career, Tsai was an expert at creating narrative. In Taiwan, he wrote and directed critical and commercial successes for the small screen. These works prove that he does not lack the capacity for narrative logic: he once relied on this to make a living. Yet, in 1990, when he ran into Xiao Kang in Taipei and invited him to act in the television film *Boys*, Tsai's filmmaking began to connect his queer life and the development of his cinema. In 1992, Tsai cast Xiao Kang as the male lead in *Rebels of the Neon God*. From this point on, the narrative logic of Tsai's cinema began to change, developing an inescapable dialectical relationship between Xiao Kang's body and life. According to Tsai, "all of this is myself,"[10] implying that he is using the screen as a mirror to engage with issues of recognition, narcissism, and self-misrecognition generated in regard to the Xiao Kang in the mirror. Tsai has said that creation is a type of "self-searching."[11] While Tsai expresses himself through the Xiao Kang in the mirror, he also tries to conceal his regard for Xiao Kang as an object of desire. Through this conception of film as personal expression, Tsai found a connection also to the films of François Truffaut, in which "it feels like life and film combine as one."[12]

Truffaut inspired Tsai to pursue his own "auteur cinema." As Truffaut's spiritual father, André Bazin, points out, "the auteur is a subject to himself; whatever the scenario, he always tells the same story, or, in case the word 'story' is confusing, let's say he has the same attitude and passes the same moral judgments on the action and on the characters."[13] Tsai's films always pass the same moral judgments on Xiao Kang; the changes the character Xiao Kang undergoes constitute the narrative theme of Tsai's films. Tsai believes in the filmmaker as auteur. In a recent interview, he asserted, "You must believe there is an author behind every film."[14] Tsai Ming-liang has come to represent the influence of French New Wave cinema on Taiwanese cinema.[15] Tsai has criticized Hollywood's monopoly over the global cinema market. This marks an important break with his auteur influences. Although Truffaut placed his own "auteur cinema" in opposition to mainstream French cinema, he and others saw no contradiction in rejecting the French filmmaking establishment while embracing Hollywood. Rather, they believed that auteurs existed in American cinema.[16] Thus, we see Tsai has not simply imported auteur theory but has carefully adopted key aspects of it.

The two main artistic features of Tsai's films—auteur theory and corporeality—have both been disputed. Therefore, I will examine the corporeality and auteur theory of Tsai's oeuvre in the context of the mirror experience, thus rethinking the question of whether Tsai's films create a closet for Chinese queers or construct a queer performativity belonging exclusively to Chinese people. Is Xiao Kang's family a queer family? I will also rethink the mirror relationship between Tsai and Xiao Kang from the perspectives of both men. In addition, the scope of intertextuality[17] of Tsai's oeuvre must be extended to cover Taiwanese TV dramas and television films,[18] dramatic works, and early literary creations in Kuching. It is this intertextual base that has fashioned the "self-expression" of the auteur director that Tsai is today.

The Subaltern Body: From Malaysia to Taiwan

Tsai, who is a Malaysian national, has always been content with his diaspora status. He says that he has always moved around: he has described himself as "drifting like a subaltern with no fixed abode."[19] Tsai's younger years were spent in Kuching, Malaysia. Before he went to Taiwan for study, he published many works under the pen name "Momo" (默默). The works include essays, fiction, modern Chinese poetry, theater scripts, and radio plays.[20] He was born into a poor family, and this class background informed his perspective in selecting materials that were observations of the daily life of subaltern groups. This laid the foundation for his film work and its focus on subaltern groups, a focus he would explore through traditional means as well as through sound and a specific gaze.

From a young age, Tsai helped his family make and sell noodles. In his essay "Yizun mingyue gumiao qian" ("A goblet of moonlight in front of the ancient temple"), Tsai and his parents are described clearing up the stall in silence. However, they could still hear the noise of the city all around them. In later films, he places great importance on sound recording and includes city noise as "resources of diegetic sound." Meanwhile, he reduces non-diegetic sound to a minimum: much of the time there is no music at all. In addition, Tsai's early poetry reveals a gaze fixated on male subalterns. For example, in *Zaijian Yulang* (*Goodbye, Fisherman*), Tsai describes a reunion with a school friend from a fishing family. Tsai gazes emotionally at his friend's tanned face and bright eyes. This gaze is the prototype of Xiao Kang's gaze at the young men in *Rebels of the Neon God* and *Vive L'Amour*. This desiring gaze also crosses racial boundaries. The poems *Linzhong de niao jiao* (*The bird in the wood calls*) and *Wo de Ludaya de pengyou* (*My Bidayu friend*) are a gift to a Bidayu friend who used to work for Tsai's family. Tsai's middle school Chinese teacher and renowned poet Tian Si believes that the Bidayu friend is the model for a radio play, *Bayue de yueliang* (*August Moon*), which touches on the issue of interracial marriage. These interracial gazes of desire are replayed in Tsai's later film *I Don't Want to Sleep Alone*.

His early work also shows a concern over the decline of the cinema. His essay "Zheyang de yichang dianying" ("A showing like this") describes a scene in a cinema where he is the only person watching the film. He remembers the crowded cinemas of his youth, and the solitude of spectatorship later in life. Tsai revisits such instances of isolated cinema attendance in *Goodbye, Dragon Inn* and the short *It's a Dream*. Tsai's 1989 television film, *All Corners of the World*, ensured his fame in Taiwan. Its tight narrative depicts a subaltern family struggling to survive as ticket touts in Taipei's Ximending district at the end of the 1980s, against the rapid decline of "film culture." The film has a clear narrative and reveals the trials of Taipei's subalterns using neorealism. It primarily focuses on a brother and sister who have grown up in a subaltern family—Meixue and Ah Tong. In the film, the violence of society and the violence of the family coexist. The theme of domestic violence reappears in Tsai's latest film, *Stray Dogs*. In his 1991 television film, *Give Me a Home*, Tsai depicts the living conditions of Third World construction workers. They build houses but are unable to buy houses themselves, living temporarily on site instead. The attention to the plight of laborers surfaces in his later film *I Don't Want to Sleep Alone*.

From his early literary works to *All Corners of the World*, *Give Me a Home*, and *I Don't Want to Sleep Alone*, Tsai's gaze is focused on these transient subalterns. This gaze is a projection of the diaspora status of his own self-expression as a Malaysian Chinese queer made a second-class citizen by Malaysian constitutional discrimination. While Tsai worked in Taiwan, his works were often discriminated against by Taiwanese and Malaysian mainstream discourse due to their treatment of homosexuality and incest. Tsai's self-expression was made "homeless and stateless," like the migrant workers he portrays. This is evident in *I Don't Want to Sleep Alone*. The film features a roofless block. Tsai mocks this ruined block, making it "impressive, like a post-modern opera house,"[21] while the rain creates a lake at the bottom of the building. At the end of the film, the maid (Xiao Kang) and the migrant worker (Rawang) lie together on a double bed, floating across the lake. It looks like they have become a family. This appears to be a roofless queer household, but can it still be regarded as a "household"? How do they dispel their worries? Because he has no home and no state, the body has become the site of self-expression in Tsai's films.

Self-expression in Tsai's films often needs to borrow a wandering body in order to locate itself within the narrative. This is why Tsai says, "when I shoot a city, it is just like I am shooting a person,"[22] the darkness of the city and the dirt of the body often combine into one. Both lack beauty when seen through the lens, especially Kuala Lumpur, the background for *I Don't Want to Sleep Alone*. In the film, the migrant laborer Rawang seeks out a homosexual relationship with the homeless Xiao Kang. These "gendered subalterns" are repressed due to their gender, race, and class. They have no means to narrate themselves and so have never had their own agency. However, this was deemed by the Malaysian film censor to be "ugly,"[23] or, put more subtly, Tsai's films were understood as lacking an aesthetic dimension, what Peng Hsiao-yen terms "a negative aesthetic technique."[24] As Peng sees it, because Tsai believes that the only aspect of a film worthy of respect is the moving image, he abandoned narrative logic.

In fact, Tsai's films have not abandoned "narrative logic," but rather make greater use of "parallelism" than they do of linear progression. In *I Don't Want to Sleep Alone*, the first narrative focuses on Xiao Kang's corporeal body as he lies on his sick bed, while the second concentrates on the story of Xiao Kang's disembodied soul as it wanders the streets. These two "parallel" narratives discard the overreliance of film narrative logic on cause-effect relationships and the continuity of time and space. It seems Tsai does emphasize other filmic elements over narrative logic in the film; however, perhaps this "parallel" narration constitutes a different narrative logic in which the benchmark is the corporeality of the "gendered subaltern" rather than a narrative logic grounded in heteronormative, bourgeois aesthetics.

Coming Out in the Mirror: Xiao Kang and Tsai Ming-liang

In Tsai's creations, the body is "a locus for developing the false, fantasy, and desire."[25] Scenes in his films often touch on the human body and the complex sexual desires it engenders. Of these, the most controversial are homosexual elements. These are evident in Tsai's early works too, especially in the play *Fangjian li de yigui* (*The Closet in the Room*). Tsai wrote, directed, and acted in this play, and he did not deny that it is a self-portrait. He has said, "creation is always selfish—I actually do it for myself."[26] Following Tsai's claim of self-referential intimacy within this play, we can trace his shift toward corporeality in his films.

Fangjian li de yigui displays the troubles that arise between the work, ideals, and desires of a creator. It skilfully uses the interaction between the protagonist and external noise to give the play an atmosphere of heteroglossia. Internal actions and sounds are knitted together, tightly linking every layer of the narrative logic. These varied sounds create cause and effect within one space and time and create an interactive logical relationship with the protagonist. They lend variety and interest to the protagonist's actions, moods, and expressions. The play as a whole appears to be the protagonist talking to himself in an enclosed space—his room. However, it reveals the difficulties being a Chinese queer, simultaneously busy and lonely. The closet is an even more perfectly enclosed space. It is a closet within the protagonist's inner world, in which is concealed a formless being. This is a metaphor for the protagonist's id, which attempts to draw him into dialogue by constantly disrupting the protagonist's daily life.

What first draws the protagonist into dialogue is a metal chocolate box containing scraps of love letters from his boyfriend. When, in the first scene, the boyfriend phones to say that he is getting married, the distraught protagonist hangs up the phone and proceeds to rip the love letters up before putting them in this box, which he then puts in the portable closet. In addition to displaying the love (and anguish) still felt by the protagonist, this gesture subtly indicates that the only thing the protagonist can do after receiving an emotional blow is push his homosexuality further into the closet.

This chocolate box draws out the existence, but not the appearance, of the formless being in the portable closet. The formless being in the portable closet is not satisfied with the way that the protagonist hides his sexuality and protests and causes trouble by turning lights on and off. These actions lend the play a surreal yet frightening humor. The process that the protagonist goes through, from his initial fear to gradual dialogue with the formless being in the portable closet, is the process of the protagonist facing up to himself. When the formless being in the closet is treated coldly, he throws everything out of the portable closet, including the chocolate box containing the torn letters. If the protagonist treats him better, he tucks him in at night and joins him in his creative endeavors. Finally, the protagonist realizes he wants to run a theater. When he returns, excited and full of gratitude, he places a rose in the zipper of the portable closet, and in the last scene the portable closet silently flies upward. The flying closet symbolizes Tsai's attempt to break free of the troubled situation he was in as a homosexual. There is no way to be sure if the closet lifted Tsai or if Tsai lifted the closet. However, we can be certain that the queer protagonist is both unable to discard the closet and unable to come out of it.

The appearance of Xiao Kang enabled Tsai's creations to give form to the being inside the closet. According to Xiao Kang, one day in 1990 Tsai left a cinema and saw Xiao Kang outside an arcade. He walked over and asked Xiao Kang if he wanted to be an actor. Xiao Kang was surprised, as he did not think that Tsai looked like a director.[27] Remembering the event, Tsai has said that he felt that though Xiao Kang appeared harmless, he was really a "bad boy,"[28] and that Xiao Kang reminded him of his younger self. Xiao Kang was working in an illegal arcade. His job was to keep watch and warn if the police arrived. Xiao Kang existed in the lower layers of society, like the middle school student in their first collaboration, *Boys*, who extorts money from a primary school pupil. Xiao Kang passed the audition for this character, and Lu Yi-Ching was cast to play Xiao Kang's mother. Xiao Kang's family was taking shape, and Tsai began to use the screen mirror of *Boys* to regard Xiao Kang as a performance of himself. This is the beginning of the mirror relationship between the two men.

In *Boys*, Xiao Kang extorts money from a primary school student. In turn, he hands over the money he extorts to a shadowy leader who is extorting him. The agents that dispense and receive violence are not dichotomously "good" or "bad." Although Xiao Kang looks like a problem child, in the eyes of his friends he is someone who will put himself in danger for them, and in the eyes of his mother he is a young and inexperienced child. When he is with one primary school student they perform the relationship of bully and victim, but when they return home both face their parents' indifference. Since they first began collaborating, then, Tsai and Xiao Kang began to rethink the traditional heterosexual family: "Any child within the family structure may become a lonely individual that appears to be autonomous but lacks inner support."[29]

Xiao Kang is a representation of this. He escapes from this traditional heterosexual family by wandering constantly. In *Rebels of the Neon God*, Xiao Kang spends all night following Ah Ze through Ximending. In one scene, Xiao Kang discovers that Ah Ze is making love to a girl in a hotel, and in his rage he smashes Ah Ze's scooter. This violent action can be regarded as an intertextual representation of Meixue smashing the scooter of the schoolmate she has a secret crush on in *All Corners of the World*, also set in Ximending. The audience feels that when Xiao Kang smashes Ah Ze's scooter it is not purely an act of revenge but rather a summation of Xiao Kang's vain love, pursuit, and hate of Ah Ze. Here, Xiao Kang is pulled by an indeterminate desire for Ah Ze; however, in *Vive L'Amour*, the homosexual feelings Xiao Kang has for Ah Ze are openly displayed. In *Vive L'Amour*, Xiao Kang finally kisses Ah Ze in secret, washes his clothes, and eats with him. These gestures make it clear that he wants to live with him. This contrasts with the character of the estate agent, who goes to bed with Ah Ze as soon as she arrives. Critic Yao Yiwei believes "the '*amour*' of the title does not exist between men and women, but rather between male homosexuals"[30] because it is the two homosexuals who are in love and who harbor a desire to live together. However, all Xiao Kang can do is live secretly in an apartment for sale because the surrounding society cannot provide a space where queers can exist openly. They can only hide in the closet, a site that calls into question their existence.

Although Tsai's films depict homosexual acts, there has never been a main character, not even that of Xiao Kang, who has admitted his queer identity or come out in diegesis. Moreover, in these films, the queers are all especially attracted to straight men. Xiao Kang keeps his sexuality, like the formless being in *Fangjian li de yigui*, in the closet. This relation between closeted desire and active (queer) male gaze is then transferred from the play's depiction of the relation between the male protagonist and formless being in the closet to the "codependent" mirror relationship between Tsai (as director) and Xiao Kang (as character). Tsai uses the mechanism of film to give the formless being a body; the formless being is played by Xiao Kang, who takes the drama's immaterial persona out of the closet and depicts him on the screen mirror. Meanwhile, Tsai himself spends much of the time behind the screen mirror, working from inside the closet. In other words, the mechanism, the apparatus of film, has become Tsai's closet. From within the filmic mechanism of the screen mirror, Tsai gazes through his camera and through all the filmic elements at the Xiao Kang in the mirror. Tsai's self-expression is given full rein, but his queerness is hidden in the closet. Tsai's subject remains silent. It does not have its own language, and neither does the Xiao Kang in the mirror. Xiao Kang has always been a projection of Tsai's ego, an ego that clings to the mirror stage of the imagination, refusing to transition to the language stage. Therefore, the queer as subject has no means even to establish itself in the diegesis of Tsai's films, let alone come out.

Perhaps, though, to understand the full significance of this lack of subjective expression outside of the closet or in the cinematic diegesis, it is necessary to interrogate why so many Chinese queers are, like the protagonist in *Fangjian li de yigui*, unable to discard the closet. Perhaps it is necessary to ask why they can only use artistic forms to display their selves, while simultaneously remaining unable to establish queer subjectivities. Perhaps it is time to ask if these virtual artistic forms are strengthening the very real closet of Chinese life and if this relation to the queer Chinese closet functions positively or negatively for Chinese queers.

Many aspects of Chinese societies work to intangibly oppress homosexual subjectivity in a way that reflects the oppression of lesbian identity in Western societies, that is, "oppression works through the production of a domain of unthinkability and unnameability."[31] In Chinese cultures, this is the societal network centered on the heteronormative kinship structures, often labeled "family values," which constitute the foundation of the closet of the Chinese queer. Homosexuality was never expressly prohibited in law in Taiwanese society. However, this does not indicate tolerance toward homosexuality but rather illustrates how homosexuality was denied even a subcultural expression that could be imagined and honestly described. In other words, homosexual subjectivity is masked by heteronormative "family values" within the mechanisms of certain aspects of Taiwanese society. Like other subaltern groups, homosexuals cannot speak or represent themselves because their subjectivity does not exist. They can only be spoken about or represented, always in the negative, as a lack. This is also why Tsai needs the Xiao Kang in his films to project his own homosexual desires. Tsai cannot represent his queer self; he can only be represented through his complex relationship with Xiao Kang.

Historically, these social structures worked to enclose these marginal groups, to deny them legal and ethical recognition. In this way, each generation of homosexuals has been pushed into a crisis of self-identity. Since subjectivity relies so much on outside recognition, and since homosexuals under these conditions cannot gain the recognition of other people, or can only gain a distorted recognition, they share in this non-recognized or distorted identity. They see themselves through the lenses of other people, and when those others fail to recognize them, they fail to recognize themselves. In other words, the source of self-identity for Chinese queers is often not direct recognition, but rather generated in the process of consuming or creating queer art. They must rely to an excessive extent on the artistic medium as a mirror in order to project and perceive their queer desires. Therefore, Chinese queer self-expression and self-identity are always misrecognized from the start, and the queer subject is always founded on a misrecognition that renders identifying as queer outside the closet impossible.

Some queer activists have criticized Tsai for using his lens to bring queers to an audience from a negative perspective, for example through a sick body, a dark sauna, or a dirty toilet. The Hong Kong queer activist and artist Edward Lam (林奕華) questions even the existence of the expression of homosexuality in Tsai's *The River*, asking, "Is the homosexuality really necessary? Or is it just an attempt to stand out?"[32] In his critique of the film, Lam also mentions several American queer film festival coordinators who had screened *The River* and not believed that it was a film about queer people because it did not work to redeem queers or express a positive image of homosexuality. However, such criticisms misrecognize the positive significance Tsai's films have had for the Chinese queer equality movement precisely for their alterations of heteronormative expectations of restorative narratives. Tsai has responded to such criticism by excoriating the queer cinema theme of the Taiwanese Golden Horse awards, initiated by Edward Lam, for "cynically trying to win people over."[33] Tsai has stated that he believes "homosexuality should not be categorized and self-differenced."[34] Therefore, he has always fought attempts to label his films as "queer cinema." This standpoint angered Edward Lam and others involved in the awards.

The question of this label and its reverberations is inextricable from the debate over queers and the act of coming out on and off screen. The discourse of the queer rights movement in Hong Kong/Taiwan can be categorized into three factions: the "coming out faction," the "middle faction," and the "hidden faction." The "coming out faction" believes queers must "come out" because they think the closet is not an incidental result of heteronormative oppression but rather an important constituent mechanism of it.[35] Representatives of this faction include Edward Lam and Taiwan's Lin Xianxiu (林贤修). The "middle faction" argues queers must have the option to come out but that the choice of coming out, and how far to come out, remains ultimately with the individual. They contend that queers should decide for themselves which risks to take, where, when, and why.[36] This faction is invested in state sanctions for gay rights and contends that without a legal structure ensuring equality of opportunity regardless of sexuality, the "coming out" of Chinese queers remains always a "coming out" that differs in degree and kind from its Western model. It remains a coming out to a particular group or in particular interpersonal relationships only. Representatives of this faction include Taiwan University academic Chang Hsiao-hung (张小虹). They run activities for queers such as a "vote for ten dream lovers" and "rainbow lovers' week," and call upon queers to mix with heterosexual groups that support the queer movement to come out collectively and make a show in front of the media.[37]

Representatives of the "hidden faction" include Hong Kong academic Zhou Huashan (周华山). They do not deny the significance of coming out but question the coming out model and strategy of the Anglo-American queer movement. This faction believes that the coming out politics advocated by the mainstream Anglo-American queer movement is a type of confrontational identity politics unsuited to Chinese societies, with their network of interpersonal relations rooted in the heteronormative familial kinship structure and with a strong emphasis on harmony, peace, and union. They assert that "the attitude in traditional Chinese culture towards homosexuality was one of tacit tolerance, not public acceptance"[38] and that very few homosexuals have been subjected to religious and legal persecution, as have those in the United Kingdom and United States. Therefore, to forcibly import Anglo-American "coming out politics" into Chinese societies would be a "rank colonization of desire,"[39] and would only succeed in forcing Chinese queers out of their societal networks, leaving them alone, isolated, and vulnerable because, they insist, in addition to losing their cultural identity, Chinese queers who forcibly come out of the closet may lose their jobs and their living spaces.

Many Taiwanese queer academics believe that since the 1990s the Taiwanese queer movement has been stuck rerunning the debate over whether queers should come out or not. The Taiwanese queer movement in the 1990s fundamentally happened in the mainstream media, and although this movement changed the image and cultural standing of queers on screen, this newly gained recognition has not transferred to off-screen experiences. The majority of queers in Taiwan remain unwilling to come out publicly.[40] Tsai, whenever he discusses queer issues in the media, stresses that homosexuals do not need to openly expound their identities; they do not need to write "homosexual" on their foreheads. He explains that though he has his confusion, shame, emotional baggage, misgivings, and societal pressure, he reveals his repressed emotions in his films.[41] To a certain extent, then, Tsai's words and actions suggest his sympathies lie with the "hidden faction" within the queer community.

The queer characters in Tsai's films often appear within a "sexual ghetto." It is as if his queers can only interact in these segregated sexual locations. This constitutes the focus of the debate between the "coming out faction" and the "hidden faction." As the "coming out faction" sees it, these representations can only add to the stereotyped impression of queers, push more queers into the closet, and, in the end, prove unhelpful in the fight for queer rights. However, the "hidden faction" claims that it is impossible to insist that every queer have a positive image: heterosexuals also have their dissolute and pessimistic side. Why not permit these queers to represent their way of life? The "coming out faction" holds that all queers have a duty to devote themselves to the queer rights movement, precisely the "impossible task" that the "hidden faction" refuses to take on. Mainstream discourse in the Anglo-American queer rights movement advocates that gay people must come out. Coming out has several advantages: it can keep homosexuals from being lonely, allow them to seek more support, solve the problem of forced marriages, and change the stereotypes society holds about homosexuals.[42] In addition to this, legal practicalities dictate that homosexuals be willing to have their identity counted. The law cannot give equal rights to a group of people who stay in the closet.

African-American scholar Anthony Appiah argues that there are an infinite number of gay and black modes of behavior.[43] However, the "politics of recognition" advocated by the queer rights movement often finds itself in a self-limiting predicament, implying a binary distinction between supporters and opponents. It forces various groups within the subaltern to move toward a simplified categorical demand—that they rally under the conception that there is only one race to be recognized, only one gendered, sexed body to be acknowledged—because of the pressing necessity for political activation. They argue that minorities must deny their own differences within differences and present a united front in order to make political gains. This demand makes it difficult for those who want to treat their skin color and gendered, sexed body as personal dimensions of the self.[44] If you were to force a black homosexual to choose between the closed world of the closet and a world of liberated homosexuals, they may well choose the latter option. However, their greater desire might be that they not be obligated to make any choice at all, or that more options be provided. Appiah believes that it is not sufficient to dwell on the right of homosexuals to come out: rights and dignity equal to that which heterosexuals enjoy under the law must be obtained.[45]

There is a radical component to the queer equality movement among Chinese people in Asia that advocates coming out. However, there are many more Chinese queers who are unable to come out because they must face both the oppression of heteronormative "family values" politics and the collective discrimination against queers practiced by state mechanisms and societal institutions. "Family values" are often the most difficult obstacle preventing Chinese queers from coming out. Chinese family values and culture place a high value on compromise and harmony, two mobilizations of the virtue ethics of the Confucian *Doctrine of the Mean*. Crucially, this system emphasizes moderation, instilling a reluctance to take a radical line. However, Lu Xun believes that Chinese people love negotiation and compromise, asserting, "without a radical proposition, they will not agree to even the mildest of reforms."[46] Though Lu Xun's remarks are not about homosexuality, they can provide a basis for rethinking the coming out debate among Chinese queers. They may even provide a practical incentive to rethinking the deployment of radicalism within Chinese queer communities, where advocating coming out might be seen as a necessary rhetorical strategy provoking negotiation. However, how this rhetoric can be translated into direct, recognizable actions in the queer community, and to what extent, continues to depend primarily on how individual queers tackle family values politics.

In one interview, Tsai Ming-liang has indicated that he finds it difficult to accept the heteronormative Chinese family-first values system: "This outlook, in which you respect your father and venerate the family, means that marriage becomes a duty through which the family line is continued. I think that in my films I have seriously questioned this value system."[47] How does Tsai question Chinese family values? He directly engages in presentations of a leaking, unhealthy household. Tsai deploys the family, alternative versions of the family, in response to "family values." He silently and symbolically removes the roof of this household—Xiao Kang's household—and cinematically queers it with water imagery. *The River* demonstrates this practice.

In *The River*, Xiao Kang runs into a female school friend and makes love to her in a hotel. Afterward, his neck suddenly becomes crooked and painful. This pain does not respond to treatment. However, in the final scene Xiao Kang stands on the balcony looking at the sky, suggesting that he has found relief. We do not know whether his relief is due to father and son coming out in the gay sauna the previous evening—when the father slapped Xiao Kang, did that twist his neck back into position? Or was it in the earlier scene where the father cuddled Xiao Kang that Xiao Kang was released from his pain? On screen, we see all that this father and son do in the dark, while the mother waits alone in a leaking house. Some scholars believe that Tsai uses the scene in which father and son make love in the sauna to arouse the deepest dread of Confucian culture—the dread of having no descendants.[48] They also see this depiction infringing on "the political correctness of the 'family' in modern society."[49] Chang Hsiao-hung believes *The River* has provided our cultural symbolic language with an imaginative possibility beyond the "patricide" of Freudian theory: "This time, when he runs into his father, he did not kill him, but made love with him."[50] Thus, Tsai reconstructs the family to resist Confucian and non-Confucian expectations, queering the relation between them.

Tsai's moral outlook has repeatedly created trouble for his works and put him under a lot of pressure. When he made *The River*, he explained that he did so "in order to have a dialogue with myself…this was the first time I truly felt the pleasure of creation."[51] Over a decade later, he represented another incestuous scene in *Face*. This time, Xiao Kang fondles his dying mother's vagina as he stands in his flooded house. Compared to the realistic representation of father-son incest in *The River*, the mother-son incest here appears somewhat surreal. Tsai deploys many mirrors throughout *Face*, and much of the imagery is constructed through reflections. As Tsai has explained, he converted to Buddhism and began to see human life as "flowers in the mirror, the moon on water" (镜花水月), a Buddhist saying emphasizing impermanence. In recent years, Tsai has also begun to regard his films as "flowers in the mirror, the moon on water," and hopes that his audiences can watch and appreciate *Face* from this Buddhist worldview: "Everything you have seen is not real, it's a reflection of something, it's an illusion."[52] The interviewer described the Tsai of today as a "mirror man," his films reflect and reflect on the relation between desire and corporeality in their displays. This description aptly sums up Tsai's cinephilia for Xiao Kang, founded on the basis of the mirror experience. The mirror relationship between Tsai and Xiao Kang over all these years is an illusion, but an illusion that provokes.

Throughout Tsai's films, Xiao Kang's household is often presented as a strange, eternally incomplete, forever water-laden household. Water is always leaking in, as if Tsai has torn the roof off, embodying but denuding desire, leaving naked desire facing the heavens. It is a roofless queer household, and Tsai periodically directs a fixed cast of actors to repeat the actions of eating, drinking, and defecating within it. Here Tsai's removal of the roof is a corporeal performance of behavioral art. In the end, Tsai demands nothing in his films. He issues no call to arms in the public debate. He even little allows his queer characters to come out at home, like ordinary queers. These queers can only appear in places such as the sauna in *The River*, the men's toilet in *What Time Is It There?*, the dark corners of the abandoned cinema in *Goodbye, Dragon Inn*, or the bushes at night, as in *Face*. Can the appearance of homosexuals in such locations, in such situations be regarded as a "coming out"?

Zhou Huashan has suggested three features that mark the more successful cases of coming out in China, Hong Kong, and Taiwan. These features include non-conflictual harmonious relationships, non-declarative practical everyday acts, and a healthy personality that is not centered on sexuality.[53] Zhou defines "successful coming out" as "honestly and openly constructing a queer identity, without denying or destroying one's familial or cultural identity, and being able to maintain harmonious and intimate familial relationships."[54] Zhou reconceptualizes "coming out" using the "family first" outlook of certain aspects of Chinese culture, differentiating it from certain Anglo-American queer "coming out" practices that focus more on linking coming out with sexual liberation. Therefore, these "successful cases" of coming out have been criticized for being too hesitant and perfectionist as they hold to more traditional identity and kinship structures. Critics hold that this type of "coming out" is not a coming out at all, but rather indicative of the way in which the homophobia contained in the reticent poetics of the political rhetoric of Chinese family values often causes the "coming out" of Chinese queers to ultimately transmute into a hiding away.[55]

Tsai has often stated publicly that he regards Xiao Kang as "family," and that Xiao Kang's silence and the way he smokes is very similar to Tsai's late father. Reciprocally, in an interview, Xiao Kang has also indicated that Tsai is like "family" to him because Tsai is good at looking after people and is like his "second mother."[56] It is apparent, then, that the intimate mirror relationship between Tsai and Xiao Kang has seeped from their on-screen relationship into their off-screen relationship. Off screen, Tsai and Xiao Kang run a café together. When Xiao Kang's mother was interviewed on television, she praised Tsai's cooking. When at home, he cooks for Xiao Kang's mother and Xiao Kang. She also commended Tsai for being good at looking after people and for his continued filial piety. In 2009, the three were invited onto a TV program, and all three regarded each other as "family." If this is Tsai's "successful coming out," then it does indeed match Zhou's definition. Tsai lives harmoniously with Xiao Kang and Xiao Kang's mother in the Xiao Kang family; he cooks for them and cares for their well-being, but he does not talk about sex in front of them. Tsai, then, appears to have "successfully come out" according to the "family values" model. Yet, at the end of the program, Xiao Kang revealed that he has had girlfriends and that he has been with his current girlfriend for six years. Xiao Kang's mother joined in to say that she has been pressing Xiao Kang to get married for ten years now. The program finally revealed that Tsai's "successful coming out" is actually a type of hiding away. The intimate relationship between Tsai and Xiao Kang has been marginalized in the reticent rhetoric of Chinese family values. If there is any "successful coming out" to speak of, what has he come out as? He has come out as a member of Xiao Kang's "family."

In 2003 Tsai was asked if Xiao Kang was his "*tongzhi* lover." Tsai answered, "I can only say that we are more than friends but absolutely not lovers! If you must say that we are *tongzhi*, then this is only true in the sense that we are soulmates (志同道合)[57] in life and in art!"[58] This answer calls into question any direct physical relationship between the two men. In a 2007 interview, Tsai again admitted that "Xiao Kang is my emotional focus point.... Xiao Kang and I have long sublimated into family."[59] The actress Yang Kuei-Mei, who has often collaborated with Tsai, has said that the feelings between Tsai and Xiao Kang were complex, resembling father and son, brothers, and "buddies." She has also called upon people not to think of their relationship in narrow terms, strongly implying that they are not gay lovers: "They are like family."[60] However, later, when at the 2013 Venice International Film Festival, a reporter asked Tsai to define his relationship with Xiao Kang, Tsai replied, "There is no definition. We are life partners."[61] The Chinese media interpreted this statement as Tsai finally coming out of the closet.[62]

Xiao Kang's fuller statements add another layer to the question of Tsai's coming out. In 2010, when I interviewed Xiao Kang, the actor stressed that he was not a *tongzhi* and revealed that he resists playing *tongzhi* in Tsai's films. Every time filming ends, he tells Tsai that he will not play a *tongzhi* again. He explained that he has nothing against homosexuals. However, often playing a *tongzhi* has led to him being typecast and provoked other directors not to ask him to be in their films. Xiao Kang has also said that many women mistakenly believe him to be a *tongzhi* and so refuse his advances. In this interview, Xiao Kang also revealed that if he met the right woman, he would consider marriage.[63] These statements suggest scholars reverse their claims that "Lee [Xiao Kang] is identified with the character he...

plays."⁶⁴ What we may actually see when we see Xiao Kang's characters in Tsai's films is Xiao Kang performing the queer Tsai's self and not the straight Xiao Kang's self.

Xiao Kang believes that "these times created the type of relationship we have."⁶⁵ Tsai became Xiao Kang's "very good friend" and "working partner"⁶⁶ because of the poor performance of the Taiwanese film industry. As he gets few opportunities to act in films by other directors, Xiao Kang has no option but to repeatedly appear in Tsai's films. Moreover, as an overseas Chinese person from Malaysia, Tsai was required to find a Taiwanese partner when setting up his film company in Taiwan. Xiao Kang was perfectly placed to provide help and cooperation.

Therefore, to label this on- and off-screen queer family "Xiao Kang's household" is a protective act. The Xiao Kang household in Tsai's films is an attempt to cover Tsai's queer identity and shield his ideals of homosexual love. Tsai tries to redeem his sexuality through his films. In other words, because his queer identity remains closeted off screen, Tsai must consistently use the mechanism of film to make Xiao Kang become queer on screen. He adopts auteur practices to legitimize the process of creating Xiao Kang's queer body while simultaneously using the creation of the "Xiao Kang household" in the films to establish an emotional home for himself while also remaining a permanent member of the Malaysian diaspora in Taiwan.

Auteur Theory: Tsai Ming-liang's Closet

No matter whether auteur theory is regarded as a filmmaking practice, as Truffaut thought in the beginning, or, as Andrew Sarris later held, "the wholeness of art and the artist,"⁶⁷ critics have always questioned its worship of the auteur director and its exaggeration of the director's role in the filmmaking process. Those who question auteur theory claim this approach to understanding the cinema neglects the collaborative nature of the filmmaking process and assert that directors are not untrammeled artists. These critics continue to remind us that film directors are surrounded by the "Babel-like buzz of technicians, cameras, and lights of the 'happening' which is the ordinary film shoot."⁶⁸ Auteur theory underestimates the influence of the film production system and environment on the auteur.

Nobody would deny that Xiao Kang has constrained Tsai. However, Tsai is evidently willing to be closeted by Xiao Kang. Tsai's self-expression has always been limited by his mirror relationship with Xiao Kang. Their relationship forms a closet that refuses entry to language and understanding, and so narrative logic must be minimized. The majority of audiences must rely on Tsai's extra-diegetic descriptions for their understanding of his films. These descriptions include the large amount of interviews with the media and academics that Tsai has agreed to undertake over the years.

Throughout these interviews, Tsai has made ambiguous confessions about Xiao Kang, generating a "mirror experience" when his audiences view the gayness and queerness depicted through Xiao Kang's body on screen. They experience an overlap and confusion between the imaginary and the real in regard to this display. This experience is one of the unique visual pleasures of Tsai's auteur films. To use Michael Lawrence's words, "Xiao Kang thus functions metonymically for the cinema of Tsai: the corporeal being of Xiao Kang is integral to this director's corpus."[69] It is so integral that Tsai demands audiences gaze at Xiao Kang's body in film after film, and when audiences gaze at Xiao Kang's body on screen, they share in the collective pleasure generated by Tsai's gaze at Xiao Kang's body through the camera, connecting desire and corporeality. These collective pleasures reconfirm the continuous development of Tsai as an auteur director, while filmic mechanisms succeed in instilling a type of "cinephilia" toward this "queer family" in the audience. In other words, Tsai's cinephilic regard for the body of Xiao Kang and the audience's cinephilic regard for Xiao Kang's household are both built on the foundation of the mirror experience as Tsai's films transfer his desire to his audiences.

With these mirror experiences, Tsai does not need to come out off screen as he continues to misrecognize straight men as himself. Through his filmmaking practices, he mimics and performs the straight man Xiao Kang. Tsai's auteur practices are both his closet and an alliance between queer men and straight men. On screen, the straight Xiao Kang is performing Tsai's queer identity. Therefore, Xiao Kang does not need to come out, except as a straight man. In the final analysis, Tsai's auteur films tell the story of a Chinese family and the need for a straight man and a gay man to form an alliance in order to live together harmoniously. The two men use the mirror of the screen to perform each other's selves. In his auteur films, Tsai has found his straight partner. His straight and masculine partner is Xiao Kang—Tsai's flower in the mirror, his moon in the water.

The image of the straight man Xiao Kang, who gives people the impression that he has fallen in love with performing gay men, exists only in Tsai's films. As in Tsai's films, many Chinese homosexuals end up lonely in life because Xiao Kang remains a projection. What can a Chinese queer come out as? Certain ethical and cultural aspects of the Chinese family do not allow an individual to exist and come out alone. If a suitable life partner can only live on the screen/mirror, then Chinese queers will only be able to come out in relation to that projection/reflection. As long as this mirror relationship between gay men and straight men continues to be widely maintained by a certain gay discourse, Chinese queer men will be forced to remain in the closet. Tsai Ming-liang's words, actions, and works have generated a closet for Chinese queers through the mirror experience. This generation is also a type of queer performativity that belongs exclusively to Chinese queers. Due to a combination of the break between self-expression and narrative logic and external factors, the unique queer performativity of these Chinese people can only be carried out in the screen-mirror. It ends in the "reticent poetics" of Chinese family value politics.

Note: This work was supported by a Nanyang Technological University HSS Start-Up Grant entitled, The Independent Chinese Film Culture in Singapore and Malaysia: A discussion of the impacts Tsai Ming Liang's Auteur Film Making Mode had on successors. I'm extremely grateful to Brian for his insightful comments on earlier drafts of this paper. My deepest gratitude goes to all those who made this paper possible, especially to Ewan and Richard Liu. Lastly, I would like to extend my gratitude to Yeo Min-hui and Lim Ding-han for proofreading assistance.

Notes

1. "Cai Mingliang tan Li Kangsheng: Rending zhongshen banlü," Taiwan *Zhongshi dianzi bao*, September 6, 2013, http://showbiz.chinatimes.com/2009Cti/Channel/Showbiz/showbiz-news-cnt/0,5020,130511+172013090600569,00.html.
2. Michael Lawrence, "Lee Kang-sheng: Non-professional Star," in *Chinese Film Stars*, ed. Mary Farquhar and Yingjin Zhang (New York: Routledge, 2010), 152.
3. In Chinese academia, the Chinese tongzhi (同志) means both "gay" and "queer." For interviews with Tsai and Xiao Kang, see Taiwanese television program *Shen Chunhua Life Show*, "Zai Luofu Gong kanjian ta de lian: Cai Mingliang," broadcast on September 21, 2007, featuring Tsai Ming-liang, Xiao Kang, and Xiao Kang's mother.
4. From a one-and-a-half-hour recorded interview conducted by the author with Lee Kang-sheng in Taipei, June, 21, 2010.
5. Peng Hsiao-yen, "*Auteurism* and Taiwan New Cinema," *Journal of Theater Studies*, 9 (January 2012): 144–45.
6. Ibid., 140.
7. Ibid.
8. Ibid.
9. Judith Butler, "Imitation and Gender Insubordination," in *The Judith Butler Reader*, ed. Sara Salih (Malden, MA: Blackwell Publishing, 2004), 124.
10. Daniele Riviere, *Dingwei: Yu Cai Mingliang de fangtan* (Taipei: Yuanliu), 72.
11. Lim Song-hwee and Hee Wai-siam, "Ni bixu xiangxin dianying you yige zuozhe: zai lutedan fangtan Cai Mingliang," *Dianying Xinshang* 147 (April-June 2011): 73.
12. Chen Baoxu interviewing Tsai Ming-liang, "*Aiqing wansui* wan wan sui!," collected in Tsai Ming-liang, et al., *Aiqing wansui* (Taipei: Wanxiang Tushu, 1994), 200.
13. Andre Bazin, "On the *politique des auteurs*," in *Cahiers du Cinema: The 1950s. Neo-Realism, Hollywood, New Wave*, ed. Jim Hillier (Cambridge, MA: Harvard University Press), 255.

14 Lim Song-hwee and Hee Wai-siam, "You Must Believe There Is an Author behind Every Film: An Interview with Tsai Ming-liang," *Journal of Chinese Cinemas* 5, no. 2 (2011): 181–91.
15 For context, refer to Peng Hsiao-yen, "*Auteurism* and Taiwan New Cinema," 133–39.
16 David Bordwell and Kristin Thompson, *Film Art: An Introduction* (New York: McGraw-Hill, 2008), 461.
17 For a discussion of the intertextuality of Tsai's auteur films, see Lim Song Hwee, "Positioning Auteur Theory in Chinese Cinemas Studies: Intratextuality, Intertextuality and Paratextuality in the Films of Tsai Ming-liang," *Journal of Chinese Cinemas* 1, no. 3 (2007): 229–36.
18 The writer thanks Tsai Ming-liang's producer Wang Cong for providing copies of video relating to Tsai's early-period television films.
19 Tsai Ming-liang, "Wo meiyou wangzhi," *Chao Foon* 489 (2002): 52–53.
20 See the appendix "Cai Mingliang's literary works published in *Literary Weekly*," Tian, Si. "Wenxue Cai Mingliang," Chao Foon, 489(December 2002): 36–37.
21 See Tsai Ming-liang, "Director's Notes," in *I Don't Want to Sleep Alone* (DVD), Taipei: Homegreen Films, 2006.
22 Riviere, *Dingwei*, 79.
23 Chen Huisi, "Dianjianju chajin Cai Mingliang *Hei yanquan*-zuiming: fanying choulou mian buli lüyounian," *Duli Xinwen Zaixian*, February 26, 2007, http://www.merdekareview.com/news.php?n=3537.
24 Peng Hsiao-yen, "*Haijiao tianya*: yiwai de chenggong? Huigu Taiwan xin dianying," *Dianying xinshang xuekan* 142 (January–March 2010): 131.
25 Riviere, *Dingwei*, 52.
26 Wen Tianxiang, *Guangying dingge: Cai Mingliang de xinling changyu* (Taipei: Hengxing guoji wenhua, 2002), 213.
27 From a recorded interview with Lee Kang-sheng in Taipei, June 21, 2010.
28 See *Shen Chunhua Life Show*.
29 Wen Tianxiang, *Guangying dingge*, 63.
30 Yao Yiwei, "Yibu meiyou jia de dianying: wo kan *Aiqing wansui*," *Sinchew Daily* (Malaysia), July 10, 1994.
31 Butler, "Imitation and Gender Insubordination," 126.
32 Edward Lam, "Xiao tuanyuan-guancha *Heliu*," Hong Kong *Mingpao*, April 27, 1997, edition D1.

33 See Chen Baoxu, "Yuwang, yapo, bengjie de shengming-fang Cai Mingliang," in *Heliu*, ed. Jiao Xiongping (Taipei: Huangguan, 1997), 25.
34 Ibid.
35 Chu, Wei-cheng Raymond, "Taiwan tongzhi yundong de houzhimin sikao: lun 'xianshen' wenti," in *Pipan de xing zhengzhi: Tai she xingbie yu tongzhi duben*, ed. Chu Wei-cheng Raymond (Taipei: Taiwan shehui yanjiu zazhishe, 2008), 197.
36 Wang Haowei, "Buyao jiao chu yaokongqi: tongzhi yao you 'xianshen' zizhuquan," Taiwan *Saodong* 3 (January 1997): 53.
37 Chang, Hsiao-hung, "Taiwan Queer Valentines," in *Trajectories: Inter-Asia Cultural Studies*, ed. Kuan-hsing Chen (London: Routledge, 1998), 257–67.
38 Zhou Huashan, *Houzhimin tongzhi* (Hong Kong: Xianggang tongzhi yanjiushe, 1997), 384.
39 Ibid., 370.
40 Chu, Wei-cheng Raymond, "Tongzhi. Taiwan: Xing gongmin, guozu jiangou huo gongmin shehui," *Nüxue xuezhi* 15 (May 2003): 117–18.
41 See *Shen Chunhua Life Show*.
42 Janis S. Bohan, *Psychology and Sexual Orientation: Coming to Terms* (New York: Routledge, 1996), 115.
43 K. Anthony Appiah, "Identity, Authenticity, Survival—Multicultural Societies and Social Reproduction," in *Multiculturalism: Examining the Politics of Recognition*, ed. Amy Gutmann (Princeton, NJ: Princeton University Press, 1994), 159.
44 Ibid., 163.
45 Ibid., 162–63.
46 Lu Xun, "Wusheng de Zhongguo," in *Lu Xun Quanji*, vol. 4 (Beijing: Renmin wenxue chubanshe, 1973), 25–26.
47 Riviere, *Dingwei*, 62.
48 Chang Hsiao-hung, *Guaitai jiating luomanshi* (Taipei: Shibao, 2000), 131.
49 Wang Molin, "Bei bianchu de jiashen: Cai Mingliang dianying zhong de fu yu zi," *Dianying Xinshang* (March 2002): 71–75.
50 Chang Hsiao-hung, *Guaitai jiating luomanshi*, 137.
51 Riviere, *Dingwei*, 73–75.
52 Alain Devraux, "Mirror Man," *Daily Tiger* 5, 39th International Film Festival Rotterdam (February 1, 2010): 1.
53 Zhou Huashan, *Houzhimin tongzhi*, 387–92.

54 Ibid., 386–87.
55 For relevant analysis, please see Liu Jen-peng and Ding Nai-fei, "Reticent Poetics, Queer Politics," in *The Inter-Asia Cultural Studies Reader*, ed. Kuan-Hsing Chen (New York: Routledge, 2007), 395–424.
56 See *Shen Chunhua Life Show*.
57 A play on words—the Chinese idiom meaning "soulmates"（志同道合）contains the characters used in *tongzhi*（同志）but in reverse order.
58 Jiang Huifen, "Cai Mingliang, Li Kangsheng 'tongzhi qing' gaobai," *TVBS* (April 17, 2003), http://www.tvbs.com.tw/news/news_list.asp?no=jcw62020030417021012.
59 See *Shen Chunhua Life Show*.
60 Xu Jinrong, Ye Wanru, Zheng Weibo, "Cai Mingliang ci sheng bu hun, qing xi Li Kangsheng," Taiwan *Pingguo ribao* (March 25, 2007), http://ent.appledaily.com.tw/enews/article/entertainment/20070325/3343378/.
61 "Cai Mingliang *Jiaoyou* rang 22 nian zhi'ai luniao 'zhongshen banlü' gaobai," Taiwan *Pingguo ribao* (September 6, 2013), http://ent.appledaily.com.tw/enews/article/entertainment/20130906/35274422/.
62 "Cai Mingliang chugui qingding nanyan yancheng: women shi zhongshen banlü," China *Fenghuang wang (Shaanxi)* (September 6, 2013), http://sn.ifeng.com/yulepindao/yulequan/detail_2013_09/06/1195897_0.shtml.
63 From a recorded interview with Lee Kang-sheng in Taipei, June 21, 2010.
64 Lawrence, "Lee Kang-sheng," 157.
65 From a recorded interview with Lee Kang-sheng in Taipei, June 21, 2010.
66 Ibid.
67 Sarris, Andrew, *The America Cinema: Directors and Directions, 1929-1968*. Chicago: The University of Chicago Press, 1985, 30.
68 Stam, Robert, *Film Theory: An Introduction*. Malden: Blackwell Publishing, 2000, 90.
69 Lawrence, "Lee Kang-sheng," 152.

CHAPTER 6

Thinking the Inutility:
Temporality, Affect, and Embodiment in *Useless* and *Walker*
Hongfei Liao

Abstract

This chapter engages with articulations of temporality, affect, and embodiment in two documentary films, *Useless* (*Wuyong*, 2007), directed by Jia Zhangke, and *Walker* (*Xingzhe*, 2012), directed by Tsai Ming-liang. Through this engagement, I investigate the cinematic potential of "inutility" aesthetically and politically to resist the logic of capitalism as a cycle of investment and reward. Given the cultural contexts and stylistic parameters of these films, this chapter positions them within the "alternative archive," described by Chris Berry, Lu Xinyu, and Lisa Rofel in *The New Chinese Documentary Film Movement*. Like other alternative films within this archive, these documentaries contest but do not replace the progressive narratives proffered by state-sponsored propaganda and capitalist consumerist culture, thereby resisting the economic logic of capitalism. Focusing on themes of estrangement and sacrifice, I assess the relation between character psychology and costuming in Jia's film and the relation between urban haste and the slowness of repetitive character movement in Tsai's. Through these cinematic articulations these films challenge the "representationalist" foundations and investment/reward logic of global capitalism through their articulation of "the inutile," which resists capitalist appropriation. However, I also note that, although they resist and open new sites to contest the ubiquity of global capitalism, these films still risk being reappropriated by the logic of capitalism. Thus, in the final analysis, this chapter invokes Dennis King Keenan's concept of cultural formations that "sacrifice

sacrifice" to rethink how cinema might "inutilise inutility," multiplying the politics of multiplicity and endless artistic (re)creation.

Keywords: Affect, Temporality, Costume, Movement, Jia Zhangke, Tsai Ming-liang, Documentary

In this chapter I study cinematic expressions of "inutility" by analyzing two transnational Chinese documentary films, *Useless* (2007) by Jia Zhangke and *Walker* (2012) by Tsai Ming-liang. Through this study, we see aspects of cinematic inutility related to filmic temporality, affect, embodied articulation, and political connotation. In *The New Chinese Documentary Film Movement: For the Public Record*, Chris Berry, Lu Xinyu, and Lisa Rofel propose the parameters of a new documentary movement within Chinese film. This movement emerged in the early 1990s and focuses more on individual issues, usually criticizing "both commercial and governmental pressures and constraint."[1] This new focus does not mean that the new Chinese documentary movement tries to replace commercial and governmental approaches to documentary filmmaking. Rather, the new movement suggests an "alternative archive"—namely, "private documentaries" that exist alongside the "public documentaries"—and articulates its concerns more through the voices of "the voiceless."[2] As well, in her work on Asian documentary filmmaking, Kuei-fen Chiu follows Berry, Lu, and Rofel's argument about Chinese new documentary film and finds "the similar concern with the voice of voiceless" in the case of contemporary Taiwan, South Korea, and India.[3] Following these discussions of contemporary Asian documentary movements, especially their concern with how these films document private issues and contest the progressive or representationalist narratives of state-sponsored propaganda and consumer culture, I argue that these documentaries by Jia Zhangke and Tsai Ming-liang might also be considered within the new Chinese film movement. In *Useless* and *Walker* Jia and Tsai document inutility as an "alternative archive" beyond the perspective of the mainstream documentaries produced by authoritarian and consumer culture.

From the start, documentary has played a crucial role in Jia Zhangke's filmmaking. He began his career shooting documentaries, producing several in the mid-1990s that screened at international film festivals, and almost all his feature films have been influenced by documentary practices. His 2007 film, *Useless*, focuses on the mainland Chinese haute couture brand Useless, which has since become "the blockbuster" fashion statement in the Chinese media world, in part as a result of a celebrity incident in the spring of 2013.[4] Jia's 2007 film received the Horizons Documentary Award at the Venice Film Festival and was screened at the 2007 New York Film Festival. As the second part of his "artistic trilogy,"[5] the 84-minute documentary is divided into three relatively separate sections: an examination of the life and labor of assembly line workers at a Chinese clothing factory located in a coastal city in southern China; a survey of the work and exhibitions of Ma Ke, the designer who founded the Useless brand; and an investigation into the miserable lives of the insolvent tailors who have been unable to compete with the mass production strategies of the global fashion industry and, subsequently, have been abandoned by consumer society. As with Jia's other documentaries and feature films, *Useless* has garnered a great deal of attention from film scholars and filmmakers, especially those who frequently disseminate their views online.[6] Many film critics have been especially interested in the film's articulation of bodily affect and temporality. For instance, Dan Fainaru observes that while time in *Useless* is "painfully slow," the film also contains "mesmerizing, often melancholic beauty", according to Christopher Bourne.[7] Other critics as well focus on the "painfully slow" and "mesmerising" affects of *Useless*, revealing the intense force of filmic time on our embodied experience of the cinema. For our sensory-motor schema, this slowness exceeds our habitual perception and engenders malaise or anxiety, which calls for political and philosophical analysis.

Tsai Ming-liang, the director of *Walker* (2012), is a Malaysian Chinese director usually categorized as a representative of the Taiwanese "Second New Wave." As film scholar Song Hwee Lim points out, Tsai experienced a transformation from concentrating on the "historical I" to focusing on the "private I" when he resigned from Taiwanese TV studio work and began making films expressing individual issues.[8] Similar to Jia's *Useless*, Tsai's 25-minute documentary also articulates issues of temporality, affect, and embodiment while criticizing the ubiquity of capitalist consumerism. *Walker* was commissioned by the Hong Kong International Film Festival and Youku as part of the "Beautiful 2012" program.[9] This documentary records a barefoot monk who walks as slowly and deliberately as possible through the hustle and bustle of commercial streets and bus and train stations in Hong Kong. His staggeringly slow gait, barely perceptible gestures, painfully bent head, near silence, and bright saffron red robe contrast sharply with the freneticism, congestion, noise, and cool or muted tones of the city around him. He stands out in his estrangement from this global capitalist metropolis. As Kathie Smith observes, the protracted step of the monk, "sometimes taking 30 seconds…causes those passing by to look and sometimes stare.... The mesmerizing film subversively takes on the most basic filmic aesthetics, altering expected: movement, pacing, watching and being watched, and of course finding meditated order amongst the chaos."[10] As Smith describes it, *Walker* challenges the basic premises of the institutional cinematic experience. Andréa Picard, writing in the program notes for the festival, expresses a similar sentiment. According to Picard, "A sublime and affecting viewing experience, *Walker* demands that we give ourselves over to a meditative, trance-like state."[11]

Both films, then, have been described as mesmerizing. Critics have written that they demand we give ourselves and our bodies over to them. In experiencing the extremely slowed time of both films, we are affected by both films psychologically and corporeally as they call us to experience (not just think about) their economics and politics.

Debates over the relation between temporality and affect are key concepts in critical debates about the films of Jia Zhangke. For example, Chris Berry uses the term "postsocialism" to define the temporal expression in Jia's *Xiao Wu* (1997) while he scrutinizes the effect produced by Jia's "on-the-spot realism."[12] As Berry points out, in *Xiao Wu*, Jia tries to question the progress-based narrative promoted by authoritarian propaganda. In this chapter, I extend Berry's analysis to focus on the potential aesthetic and political effects of Jia's temporality, affect, and embodiment in *Useless* (alongside Tsai's *Walker*). In terms of affect, scholars tend to maintain human-centric and representationalist outlooks in the sense that they view "affect" as a synonym for "sentiment" or "emotion." For instance, Stephen Teo observes Jia's expression of sympathy for common people as "idiosyncratic but highly affecting realism," and Yvette Bíró looks at it as an example of the director's "tender, lyrical, contemplative attitude."[13] Here, I am not arguing that their understandings of Jia's articulation of affect are wrong. However, I am claiming that such conceptions of affect privilege representable or human-centric experience and follow the same reductive logic of the capitalist system contested in my later analysis. Therefore, contrary to this accepted understanding and usage of affect, in this chapter, I argue that we take the concept (and analysis of these films) beyond human-centric and representationalist logics. By mobilizing Deleuze's theories of embodiment and affect, we can begin to think through these films in resistance to the representationalism and discipline of global capitalism.

Likewise, in terms of Tsai's work, there are also some scholars who have engaged with issues of temporality, affect, and embodiment with regard to the experience of his films. For instance, Tiago de Luca scrutinizes the affects produced by the deployment of everyday sensory experiences and considers how they "challenge and question notions of generic type, personal identity and models of normality," especially, models of "humanness itself."[14] In this analysis, de Luca problematizes the very subject of human-centric sensory representations. Following de Luca, I provide further elaboration of the aesthetic and political connotations of such representations by emphasizing their unfamiliarity rather than their familiarity. Looking at the alienating effect of the monk's eccentric, exaggerated gesture and movement, I draw out their failed sacrificial logic. Overall, scholars consider the time of Tsai's films to be non-linear, not progressive time. Some have called it "heterochronos," recalling Lyotard's reflections on "postmodernist time."[15] Others have labeled the experience of Tsai's cinematographic time as an experience of "temporal dysphoria" where protagonists and filmgoers encounter a confusion of time.[16] Here, I pursue Andrea Bachner's view on time through the theoretical perspectives of Gilles Deleuze, who emphasizes temporalities as anti-representationalist, multiple, and nonhuman events.

Deleuze offers a lens to better explore the capacities and affects of embodiment, especially with regard to the inutile bodies expressed in these two documentaries. Rephrasing Spinoza, Deleuze emphasizes the functionality of embodiment in his work by concentrating on what a body can or cannot do, rather than what a body is or is not.

According to Deleuze, "we do not even know what a body can do: in its sleep, in its drunkenness, in its efforts and resistances. To think is to learn what a non-thinking body is capable of, its capacity, its postures."[17] For Deleuze, one potential line of thinking lies in discovering what we can learn from the body—"to discover more in the body than we know and hence more in the mind than we are conscious of."[18] Through an exploration of the undisciplined body, which is the sediment of pure potential, argues Deleuze, we can locate how affect is engendered and to where it migrates.

Building on Deleuze's question of what a body can do and his conceptions of time and affect, I address two questions regarding *Useless* and *Walker* in particular, and perhaps, the "alternative archive" of the *New Chinese Documentary Film Movement* in general. First, I ask what temporalities, affects, and embodiments do these documentaries articulate and how are they articulated? Second, I ask what kind of aesthetic and political possibilities can we draw from these cinematic articulations? Emphasizing their anti-representationalist characteristics and considering their inutility as contests to global capitalism, I conceptualize the temporality, affect, and embodiment of these films, and others like them, as expressions of inutility. Instead of simply labelling this inutility as non- or anti-productive, though, I radically rethink inutility in terms of "inutilizing the inutility," highlighting the dynamic failure and estrangement in these films. This idea of "inutilizing the inutility" is inspired by Dennis King Keenan's thinking on sacrifice in *On the Question of Sacrifice*, where Keenan emphasizes the impossibility of resisting global capitalist consumerism once and for all with some final solution. Keenan stresses that it is necessary to resist global capitalism, but through dynamic, rather than finite, practices. By arguing for inutilizing the inutility, I propose we reconsider inutility as endless, inutile resistance to the ubiquitous reappropriation of global capitalism and an opening to inexhaustible artistic creations. In this sense, the inutility articulated in *Useless* and *Walker* embodies the ethic of failure, not as the human-centric possibility but as the inexhaustible multiplicity that ceaselessly resists being appropriated and encoded.

Utile and Inutile Time in *Useless*

In view of the different temporalities articulated in the film, *Useless* can be divided into three sections: the monotonous, progress-based time of assembly line workers in a Chinese clothing company in Zhuhai; the ideal artist's time of fashion designer Ma Ke, which introduces the nonhuman affect of material bodies; and the obsolete time of the insolvent tailors who cannot keep pace with the mass production model of the global garment industry. In terms of utility/inutility, we might classify the segments in the following way. While the productive time of the assembly line worker is "utile," and the abandoned time of the bankrupt tailor is "inutile," the artistic time of Ma Ke is more complicated. In one way, the artistic time of Ma Ke articulated in the defunct brand "Useless" is ostensibly inutile; yet, it remains utile as it can be further assimilated into global capitalist fashion production. It is this very threat of utility as the subtle reappropriation of inutility into the economics of capitalist representationalism that remains to be contested artistically and politically. This is the concern of inutilizing the inutile to contest appropriation.

Useless begins with a depiction of the working conditions and utile time of assembly line workers. Rather than opening with a wide-angle shot to offer a general description of the clothing factory, Jia starts with a close-up of the absorbed yet benumbed look of a male worker. The film then moves to several shots of overhead clothes hangers, while the soundtrack is filled with the monotonous noises of the machines. This stylistic strategy distinguishes Jia from many of his fifth-generation precursors whose films begin with panoramic shots that situate and contextualize their imagery in specific times and places. For example, Chen Kaige's *Yellow Earth* (1985), which has been inspirational for Jia Zhangke himself,[19] opens with a panoramic shot depicting the barren Loess Plateau (*huangtu gaoyuan*) before starting the story of the individual characters. Time and space are particular and relevant in *Yellow Earth*. It is as if they do not exist, let alone matter, in *Useless*. This factory and this work could be done anywhere, by anyone, but its effects are measured on the specific bodies deployed in the film. Jia's close-up articulates the affective and somatic state of the workers, while the ostensibly objective wide-shot aims to depict simply the larger space. This method of starting with the particular human and turning toward the general space leads Zhen Zhang to regard Jia's method as "human-orientated,"[20] as Jia's framing expresses sympathy for the mute and oppressed workers who have to sell their labor time to the global capitalists in the generic sweatshops and contribute anonymously to China's position as the largest clothing exporter in the world.[21] In displaying the hegemony that pushes workers to sacrifice their personal time to produce standardized costumes, the temporality of global capitalism deployed in this documentary is doubtless "object-orientated."[22] However, this deployment (of cutting between the specific and the general) serves as the prompt for us to consider the expansiveness of the hegemony of global capitalist time over other multiple temporalities and the manners in which we must contest and resist it, most importantly, to restore time as multiple and anti-representationalist. The opening scene also displays the benumbed look of the male worker and his monotonous work, articulating global capitalism's suppression and homogenization of the (male) body in service of capitalist production at the cost of its own multiple and creative potential.

This estranging function of global capitalist time is also vividly depicted in the third section of *Useless*, where images of two groups of artisanal workers are juxtaposed: the totally insolvent tailors who have given up their work, and the tailors who are threatened by foreseeable bankruptcy. In an interview, one insolvent tailor describes his reasons for giving up, explaining, "A hand-made suit costs at least 40 yuans, but you can buy a suit made on an assembly line for only 30 yuans. After realizing this, I thought my career is over." It is impossible for artisanal tailors to compete in terms of price with the mass-produced attire of global capitalism, at least in small town China, where the scarcity of middle-class people who can afford such consumption prohibits the artisanal investment. Thus, the tailor gives up his career. Obviously, his is not the only case, and, as the film continues to show, the time of these insolvent tailors becomes inutile because of its failure to be appropriated by the representational logic of capitalism.

The essence of utile time lies in the representationalism or reductionism of global capitalism, which remains hegemonic and merciless with regard to human subjects. In his *The Philosophy of Money*, Georg Simmel analyzes and criticizes the representationalism of capitalism. According to Simmel, "money expresses, as it were, the purely commercial element in the commercial treatment of things, just as logic represents comprehensibility with reference to comprehensible objects."[23] Thus, money (capital) functions as the minimal unit of exchange, which "frees the objects from their bondage to the mere subjectivity of the subjects and allows them to determine themselves reciprocally, by investing the economic function in them."[24] The logic of money ruptures the logic of subject-object relations, freeing objects to relate only to themselves, through money. As a result of this representationalism, every instant of a worker's time is calculated and evaluated in terms of productivity, in terms of objects produced to produce money. This is what Mary Ann Doane calls "the standardization and rationalization of time," which reduces time to the logic of exchange value or money.[25] Within this logic, anything that cannot be recycled properly into the process of exchange is inutile.

Anything that cannot be exchanged for money is useless. Besides the capitalist representation of time, this reduction logic also takes root in our daily way of seeing the world and of naturalizing our habitual recognition and identity. Hence it is also necessary to rethink and contest this reduction philosophically. The "intelligibility and simplicity"[26] of the logic of money—the logic that everything has value that can be reduced to a known quantity of a known universal—is essentially a scheme of representationalism and plays an essential role in naturalizing the reduction of the multiplicity of other values, other systems, other logics. To begin to contest representationalism, it is worth remembering Deleuze's reading of Henri Bergson on the issue of intellectual reduction. The argument is that because the brain is always "choosing now this and now that movement in response to the many it receives,"[27] without reduction we could not function intellectually. However, against this backdrop, Deleuze rereads Bergson's theory of time as a theory of anti-intentionality, where, "subjectivity is never ours, it is time, that is, the soul or the spirit, the virtual."[28] Reclaiming a Bergsonian conception of time as duration, Deleuze argues for its priority as the origin of human subjectivity, in our daily perception. In this sense, time is far from human-centric, though we get used to the linear mode of time and habitual recognition, which are subject-oriented. Thus, representationalism is naturalized and seen as a product of our intellectual capacity that overlooks the multiplicity of reality and needs to be overcome philosophically and artistically.

In the end, both utile and inutile modes of time are produced by the logic of capitalism and, thus, cannot overcome the logic of capitalism. In a similar light, as Zhen Zhang has argued, so are the "progressive individual" and the "negating subject."[29] Both conceptions of human possibility are rooted in the logic of representational capitalism; therefore, neither can transcend this logic. To further document the pervasive, relentless logic of global capitalism reappropriating these modes of time and these conceptions of human possibility, Jia considers in greater detail the insolvent tailors who populate the third segment of *Useless*. According to the film, the abandoned time of these displaced, estranged tailors, who used to spend their days working to earn their living by selling their handmade costumes, has been made inutile. However, it is later reappropriated as utile time when they are forced to become coal miners to survive. Once artisanal craftsmen who made their living through the productive expenditure of their subjective time, these men have been suppressed and homogenized into anonymous wage labor. In order to capitalize on their situation, the system recycles and reappropriates their time and their bodies, leaving nothing behind, making them utile again.

Because both can be reappropriated by capitalism, then, the relation between the utile and the inutile remains complicated. It is through this process of appropriation and reappropriation that global capitalism territorializes and reterritorializes differences, reducing the multiplicity and heterogeneity of temporalities, affects, and embodiments under the hegemonic representationalism of money (capital). Therefore, Deleuze is not only joking when he recalls the importance of "the old curse which undermines the cinema: time is money."[30] According to Deleuze, then, representationalism cannot be resisted by the logic of utility/inutility, but only by an "endlessly relaunching exchange,"[31] which would infinitely sacrifice its potential. Such an exchange is an always and forever beginning and ending, aesthetic/political task that is never done resisting the capitalist process of endless territorialization and reterritorialization.[32] Resistance must be ceaseless, for capitalism "progressively integrates the critique of capitalism into its mode of functioning, with the result that capitalism appears stronger then ever;"[33] it returns more forcefully after every challenge. Therefore, the possibility of being reappropriated makes the task of resisting capitalism an incessant mission instead of a final solution. Considering these factors, turning toward a more utile or more inutile route leads only in circles. Rather, the task at hand remains to see to the failure of both utility and inutility, to see the relation between utility and inutility radicalized through inutilizing the inutile.

Nonhuman Affect and the Material Body in Useless

After interviewing the insolvent tailors, Jia uses a long take to scrutinize the coal miners' clothes hung in the yard. The clothes are tattered and faded through repeated wear during difficult labor; they stand as images of time and bear witness to the exacerbated living conditions of the coal miners. The threadbare clothes are more than ruined apparel, of course, because they function also as reminders of the sartorial professions these men have abandoned and display the turn from utility to inutility that marks these men's bodies. For Jia, they serve as the material memory of the utile turned inutile. However, this inutile affect does more than simply confine to us, though, for affect always also points toward unrealized capacities or possibilities.

This inutility of affect also reflexively refers to the impossibility of representing embodied affect itself.[34] For Deleuze, a body is a "whole composed of parts, where these parts stand in some definite relation to one another, and has a capacity for being affected by other bodies."[35] In this respect, the essence of a body lies in what it is capable of doing or having done to it, what affects it and what it affects it engenders. Thus, because Deleuze's definition is a functional one, and different bodies can do different things and have different things done to them, it is impossible (or nearly so) to tie a body or its affects (both as affecting and affected) to one fixed perspective. Affect functions as a "transitory thought or thing that occurs prior to an idea or perception."[36] Naming an affect reduces it to its effects. Therefore, affect, as embodied, is not necessarily the reflection of a subjectivity, even if it does mark a body. Hence, it is greatly different from our notions of daily emotion or sentiment, which are based on the existence of a sensory subject. If we follow this understanding of affect, it cannot simply be reduced to the perceiving human mind or oriented body. It is the pure movement, the "in-between" bodies, "because it was never *in* any point. It was in *passage* across them all."[37] Affects do not inhabit bodies (like traits or gestures) but exist only as movements between them. In light of this extremely contingent "in-between-ness," this dimension of affect is anti-representationalist, as the representationalist perspective fails to grasp the heterogeneity and multiplicity of embodied affect. It fails to grasp it because there is nothing to grasp. In this respect, affect is nonhuman, and it is this nonhuman affect Jia articulates by filming the coal miners' mute clothes in a manner that resurrects the multiple temporalities suppressed by the estranging hegemony of capitalist-quantified temporality, as a silent contest of the relentless reappropriative logic of capitalism.

Ma Ke's motivation in naming her clothing brand is based in acknowledging its inutility, in light of its defunct design. In an interview in the second section of the film, Ma Ke explains her inspiration in designing the brand "Useless":

> I even want to bury them under the earth, to interact with nature. In other words, I'm not the one who is in charge of the whole effect but contribute something to nature. I am just the creator at the primary stage.... All I want is the affect from time and matter. I believe that there must be memory inside the matter.

Here, Ma Ke emphasizes that the real enchantment of "Useless" comes in-between, from the interaction between human creation and nonhuman nature, namely, the affect across the human mind, the clothes, and nature. Thus, she expresses explicitly the potential of the material body and the characteristics of affect.

As the exhibition in the documentary shows, the clothes seem too crude and too out of style to serve consumerist utility. No one wants to buy them. Yet, as soon as their very unfashionable appearance becomes fashionable they are reappropriated to become utile again. Though her works can be and were regarded as inutile avant-garde artistic expression, as they became fashionable, as they began to appeal to the burgeoning desires of the middle class, they became consumable fashion. Now, as an haute couture designer, Ma Ke has become a part of global capitalist production, no matter how much she disdains the costumes produced by assembly lines for their repetitive semblances and homogeneous sentiments. In this sense, the nonhuman affect presents itself, in one instance, as the estrangement between human mind and the exhibited clothes and, in another, the potential to be reappropriated by the capitalist demand of making the clothes "old" enough to be exhibited in Paris and sold during Fashion Week. It is just this possibility of being reappropriated that reminds us of the endless task of resisting the logic of capitalism in a way that leaves nothing to be reappropriated.

Untamed Time and the Estranged Body in Walker

If Jia's documentary seems to have a direction and a point, perhaps even a story to tell—through the in-depth interviews with insolvent tailors and Ma Ke that can be regarded as "on-the-spot realism"[38]—then Tsai Ming-liang's *Walker* is vastly different because it has no discernable story and little obvious direction, only a slowly walking monk to follow. In terms of its affect, *Walker* is quite "mesmerising"[39] or perhaps even intolerably tedious. What Tsai documents in this twenty-five-minute short film is quite simple: a barefooted monk in a red robe proceeds at an almost static gait through the hustle and bustle of the streets of Hong Kong with a McDonald's hamburger and a fully filled white plastic bag in his hands. It is an extremely slow film, as the protracted movements of the monk extend each step he takes over twenty-five or thirty seconds. Through long shots, the film draws our attention to the affect engendered by the contrast between the monk's slow pace and the moving urbanity that surrounds him. By way of this and other contrasts, *Walker* contests the consumerism of global capitalism.

The filmed body is at the central image in almost all Tsai's works. As Tiago de Luca observes, these bodies are usually "undisciplined, particularized and animal-like."[40] *Walker* displays a similar treatment of embodiment. Here, the image of the unhurried and unsettling pace of the barely-walking monk deploys a gesture of the untamed body distinguished by its undomesticated behavior from the hustle and bustle of hegemonic metropolitan life. With each elongated, nearly motionless step the monk's body expresses an untimely temporality, and the force of his image contrasted against those around him makes him seem out of time and place, foreign, alien. It is as if his slow movement moves him into a different time, one not quite in sync, estranged from those around him. If the "docile body,"[41] like the "quickly" moving corpus of the metropolitan, is the disciplined body, then the posture of the walking monk expresses the untimely temporality that presents itself as unpredictable and disordered. In this regard, the slowness of his embodiment contests the performances of those around him. However, his image is not only "a reminder to slow down,"[42] but also serves as a stubborn resistance to the representational logic of global capitalism because his behavior is inutile and resists becoming utile simply by moving and gesturing more slowly than anyone might have thought possible.

Near the end of the film, after we see scenes of him crossing a number of cityscapes, the monk stops in front of an iron gate and starts eating his hamburger at the same slow speed. It is interesting to observe the cultural connotations of the McDonald's hamburger, which functions as fast food in daily life and always relates to the quick speed necessary for a metropolitan: eat quickly, save time, and never be late. This quick pace is exactly what capitalism demands, in order to increase consumption and decrease production costs. In this regard, there is a sharp contrast between the slow speed represented through the monk's movements and gestures and the capitalist temporality implied in the hamburger. Juxtaposing these temporally different entities, Tsai's film plays with postmodern consumer culture and global capitalism's obsession with speed. With this strategy, *Walker* exemplifies what Michael de Certeau calls "use and tactics": though the dissident cannot leave his position, he "establishes with it a degree of plurality and creativity. By an art of being in between, he draws unexpected results from his situation."[43] As the scene shows, though, the hamburger implies the hegemonic temporality of capitalism and the monk's pace in consuming it resists its logic while amplifying its implications. Resistance does not come from abandoning the hamburger but in consuming it differently. Tsai's use of the hamburger in an unexpected way is a creative tactic to debunk the utile logic of global capitalism that undergirds our daily use of fast food. In this regard, his juxtaposition of the monk and hamburger appropriates, exposes, and mocks the appropriativeness of global capitalism.

Exposing this failure of final capitalist reappropriation, Tsai's postmodern joke contrasts the utile speed of the metropolis and the hamburger with the inutile speed of the walking monk, radically contesting the utile/inutile logic of global capitalism. Yet, capitalism can still return to re-re-reappropriate Tsai's use of the inutile (the body of the monk) and the utile (the hamburger) as elements of avant-garde resistance in *Walker*. This is the ever-present danger of capitalism. Thus, Tsai's resistance calls for an even more radical rethinking of and approach to inutility.

Radicalizing Inutility

At this point, resisting the logic of global capitalism may appear futile. As both Jia's *Useless* and Tsai's *Walker* show, almost every tendency to resist appropriation fails. For instance, Ma Ke's aims to contest the stereotype of successful fashion with her defunct design of "Useless," originally expressing her reluctance to make a utile costume; however, her resistance has been reappropriated by the capitalist logic as a new "cool" fashionable costume. In this sense, Ma Ke's "Useless" has become useful because it has become a successful brand. Another example in Jia's *Useless* is that the inutile time of insolvent tailors turns out to be the utile time of miners, revealing again the relentlessness of the reappropriation of global capitalism. As Tsai becomes more internationally recognized and his films ever-more canonized, his slow-walking monk and reappropriated McDonald's hamburger may become symbols of resistance, symbols representing resistance and, thus, reappropriable. (Even this chapter runs the risk of representing Tsai's images as images of resistance and appropriating them.) Therefore, not only does resistance seem futile but so too does the very accuracy of articulating something as inutile, since the ubiquitous logic of global capitalism assimilates and encodes everything in order to make everything utile, exchangeable, and consumable. Though Jia Zhangke's and Tsai Ming-liang's documentaries attempt to articulate affects contesting global capitalism's relentlessness, they have always already failed, as there always already has remained the possibility for capitalist consumerism to neutralize, homogenize, and reterritorialize them. This inevitability reveals the predicament in resisting the logic of capitalism. Therefore, it is necessary to call for a radical conceptualization of inutility, to overcome the danger of reappropriation. Resisting the logic of global capitalism must be a rejection of the easy satisfaction of actual political antagonism.[44] In other words, resistant resistance would have to come in the form of an endless resistance to the reappropriation of actual political agendas, and could not appear as a returning to a pre-capitalist condition or a realizable utopia.

Here, I would propose following the strategy described by Dennis King Keenan as the attempt to "sacrifice the sacrifice"[45] and radicalize our thinking on inutility to articulate a strategy to "inutilize the inutility." As its original meaning reveals, "sacrifice" sacrifices everything offered in the ritual. It is the holocaust in which all *holos* (oblation) is burned, with nothing left. Therefore, sacrifice is about absolutely giving, "to give oneself wholly to limitless abandonment."[46] In a sacrifice, one gives up everything one has to attain the authentic meaning of sacrifice, including one's own consciousness and intentionality in this action. In this case, sacrifice is already a paradox: you cannot proceed with intention. (This facet of sacrifice, of course, recalls Deleuze's rereading of Bergson with regard to time and intention.) If we sacrifice with an intention, then we are still in the cycle of economy, expecting to recycle what we have given. In that case, we still retain something to be reappropriated. That is why an absolute sacrifice sacrifices even the intention to sacrifice. Thus sacrifice is actually incompatible with itself; it is the quintessential predicament. Therefore, a radical conceptualization of inutility would strive to "inutilize the inutility." Such a conceptualization would have to have the capacity to resist the logic of capitalist representationalism through a counter-actualization, which would negate any compliance to stereotype and easy actualization. Because any attempt to simply abolish the economy of exchange would only result in catastrophe and barbarism, a radical conceptualization can only be a counter-actualized "politics of affect,"[47] which would resist any actual condition and easy satisfaction. Any singular contest would be subject to reappropriation or reterritorialization; therefore, this conceptualization would have to be a multiple, endless one, promising a future-oriented politics that is "not directed to the attainment of certain goals, the coming of fruition of ideals or plans."[48] Though it would be an endless and counter-actualized task to "inutilize the inutility," it would not be an insignificant one or one that rejects doing anything. As Keenan asserts, "rather than merely a work to be accomplished, or an action to be performed, sacrifice is experienced as a call to action that calls itself into question. It becomes a work that unworks itself in the very performance of the work."[49] In other words, inutilizing the inutility demands an endless problematizing of the actual inutility (as contest and resistance) for the purpose of endlessly relaunching more action and rethinking. In these two documentaries, temporality, affect, and embodiment are always more than representation and utilization. In this way, Jia and Tsai document these usually overlooked dimensions of inutility.

For instance, Jia's approach to documenting Ma Ke's travel to the hometown of the insolvent tailors in the final scene of *Useless* highlights the affect of this failure. Ma Ke intends to contest the stereotype of current fashion design but is herself reappropriated and becomes a successful node in global fashion production. Near the end of the film, she drives to the small inland city of Fen Yang to find workers who can produce her "Useless" line. By this point in the narrative, she has become the capitalist who deprives the insolvent tailors of their time, and her journey becomes a symbol of the representationalist logic of capitalism. Jia does not comment on her journey. He only follows her, his camera focused on the intrusiveness of her Toyota vehicle upon this locale. Ma Ke and her expensive vehicle are estranged from the small town and its people. One man stares at her vehicle with a confused and dull look for some time after Ma Ke has already passed by. Here, Jia's filmmaking is inspiring because it silently contests the reappropriation of capitalism; it articulates its resistance not through verbal expression but through embodied affect. The voiceless affect expresses itself in the form of failure, the failure to relate to, to understand the "civilized" intruder. (And here, of course, Ma Ke's big, fast, civilized Toyota functions as the inverse to the bent, slow, walking monk.) In my view, this inutility implies more possibilities of contestation than an explicit verbal expression could. *Useless* does not accuse or compliment Ma Ke. Rather, it prompts the audience to judge the scene. Hence, it maintains the potential multiple possibilities without confirming any actualized contest as the one right choice of action, and risks never completing the endless task of "inutilizing the inutility."

Likewise, in Tsai's *Walker*, the monk's embodied affect exposes the failure of being utile, which already contests daily metropolitan life because it calls for slowing down and reconsidering our actual living embodiment. However, this exposure does not overcome the problem of appropriation. The demand for "inutilizing the inutility" also calls us to problematize Tsai's filmic problematization and even our own problematization. For example, a probable route for further problematizing his cinematic expression might be to question his deployment of a monk, especially a Buddhist monk such as the one he has chosen. Depicting a monk walking at such an extremely slow speed is, of course, a reasonable artistic exaggeration; however, what further resistance demands is an interrogation of how much Tsai (and viewers?) idealize the image of the monk in opposition to the life of the metropolis. Jia's singular expression fails to articulate the diversity of the monk's other possible movements and gestures and the diverse existences of multiple monks. Therefore, it falls again into a cultural representationalism or essentialism. Furthermore, Tsai's filmic expression is radical and impressive because of its novelty. However, Tsai's contest of global consumer culture is still actually consumed by audiences and still runs the risk of being reproduced. His cinematographic approach would quickly become a fixed stereotype if it were reproduced over and over again. In this regard, we are called to continue "inutilizing the inutility" even of Tsai's film as a radically counter-actualized action.

Conclusion

Temporality, affect, and embodiment are key elements articulated in Jia Zhangke's and Tsai Ming-liang's documentaries. Through these elements, these filmmakers are able to articulate a certain resistance to the ubiquitous logic of global capitalism, which neutralizes and encodes everything it can under the principle of representationalism. One way of conceptualizing these filmic articulations is to think of them as strategic moments in inutility that resist being encoded and assimilated by the logic of capitalism. These moments, then, cinematographically contest the hegemonic temporality of global capitalism and its discipline and control of the bodies of clothing factory workers, coal miners, fashion designers, monks, and metropolitan inhabitants. By deploying such moments, these films efficiently resurrect: 1. the multiple temporalities of individuals and even nonhuman life in place of the hegemonic progress-based temporality of capitalism; 2. affect as anti-representationalist, anti-teleological, and beyond human-centric instead of representationalist, teleological, and subject-orientated sentiments; 3. embodiment as incessant assemblage and recomposition through their affecting themselves and other bodies rather than the tamed and coded body under the discipline of capitalism. In the end, of course, there can be no final overcoming or movement beyond the logic of global capitalism. Instead, inspired by Keenan's works, in this paper I radically rethink inutility as inutilizing the inutility in order not only to articulate an endless resistance to reappropriation under capitalism but also to commit to a future-orientated politics of affect and incessant artistic creation.

Notes

1. Chris Berry, Lu Xinyu, and Lisa Rofel, *The New Chinese Documentary Film Movement: For the Public Record* (Hong Kong: Hong Kong University Press, 2010), 148.
2. Ibid., 180.
3. Kuei-fen Chiu, "Afterword: Documentary Filmmaking as Ethical Production of Truth," *Concentric: Literary and Cultural Studies* 39, no. 1 (March 2013): 204.
4. In the spring of 2013, the first lady of the People's Republic of China, Liyuan Peng, chose Exception as her favourite costume while paying visits to foreign countries with her husband. This event led to an unprecedented enthusiasm on this domestic luxury. See http://ent.china.com.cn/2013-03/27/content_28370607.htm (Jiang siyuan, "The female actresses who follow the first lady and wear Exception, 2013) (accessed 10th May, 2013). It is worth noting that there are also some reports of doubt as to whether the first lady wears Exception or Useless, though Ma Ke created both lines. See http://www.forbeschina.com/review/201303/0024525.shtml (Chen Shuzhe, "Exception or Useless? The enigma to the costume of the first lady", 2013) (accessed 10th May, 2013)
5. The first section of the trilogy is *Dong* (2006), which documents a painter, and the third is still in the planning stage.
6. Here, I rely upon Internet data rather than information from more traditional published works. First of all, there are few published texts on short documentaries like *Useless* and *Walker*, thus it also affirms the necessity of doing research on them; secondly, online critique by scholars and practitioners becomes more and more influential and essential, given that it is an immediate and instant response to the burgeoning production of films.
7. Dan Fainaru, "Useless (Wu Yong)," ScreenDaily (2007), accessed 10th May 2013. http://www.screendaily.com/useless-wu-yong/4034477.article.
8. Christopher Bourne, "Jia Zhang-ke's "Useless" – 2007 New York Film Festival Review," Meniscus (2007), accessed 9th May 2013. http://www.meniscuszine.com/articles/20071106792/documentary-review-jia-zhang-kes-useless/.

9   Song Hwee Lim, "Confessing Desire: The Poetics of Tsai Ming-liang's Queer Cinema," in *Celluloid Comrade: Representations of Male Homosexuality in Contemporary Chinese Cinemas* (Honolulu: University of Hawai'i Press, 2006): 127.
10 Youku is a mainland China based website on which people can share their audio-visual files, like a Chinese version of YouTube. See http://www.youku.com/.
11 Kathie Smith, "VIFF 2012 Review: BEAUTIFUL 2012 Is a Rare, Successful Omnibus," *Twitch* (2012), http://twitchfilm.com/2012/10/viff-2012-review-beautiful-2012-is-a-rare-successful-omnibus.html.
12 Andréa Picard, "Programmer's Note," TIFF (2012). This citation can also be found in this link, accessed 31th May, 2014, http://www.liveguide.com.au/Events/823860/Artists/Walker
13 The frequently used notion of "postsocialism" is more or less indebted to Lyotard's definition of "postmodernism" as "a loss of faith in the grand narrative" of socialism (Chris Berry, "Jia Zhangke and the Temporality of Postsocialist Chinese Cinema: In the Now (and Then)," in *Futures of Chinese Cinema: Technologies and Temporalities in Chinese Screen Cultures*, ed. Olivia Khoo and Sean Metzger [London: Intellect, 2009], 115). Berry's usage of "postsocialism" is explicitly inspired by Jean-François Lyotard's "postmodernism," which indicates a doubt on the grand narrative. It also signifies a doubt on the progress-based historical narrative in the post-Maoist era and thus is a multiple and non-linear temporal articulation. For more on postsocialism, see Philippa Lovatt, "The Spectral Soundscapes of Postsocialist China in the Films of Jia Zhangke," *Screen* 53, no. 4 (Winter 2012): 418–35; Jason McGrath, *Postsocialist Modernity: Chinese Cinema, Literature and Criticism in the Market Age* (Stanford, CA: Stanford University Press, 2008); and Peter C. Pugsley, "Postsocialist Modernity: Chinese Cinema, Literature and Criticism in the Market Age," *Continuum: Journal of Media & Cultural Studies* 24, no. 2 (2010): 325–27.
14 Stephen Teo, "China with an Accent: Interview with Jia Zhangke, Director of *Platform*," *Senses of Cinema* (2001), http://sensesofcinema.com/2001/feature-articles/zhangke_interview/; Yvette Bíró, "Tender Is the Regard: *I Don't Want to Sleep Alone* and *Still Life*," *Film Quarterly* 61, no. 4 (Summer 2008): 40.
15 Tiago de Luca, "Sensory Everyday: Space, Materiality and the Body in the Films of Tsai Ming-liang," *Journal of Chinese Cinemas* 5, no. 2 (2011): 175.

16 Andrea Bachner, "Cinema as Heterochronos: Temporal Folds in the Work of Tsai Ming-liang," *Modern Chinese Literature and Culture* 19, no. 1 (Spring 2007): 60–90.
17 Fran Martin, "The European Undead: Tsai Ming-liang's Temporal Dysphoria," *Senses of Cinema* 27 (July 2003), http://www.sensesofcinema.com/2003/feature-articles/tsai_european_undead/.
18 Gilles Deleuze, *Cinema 2: The Time-Image*, trans. Hugh Tomlinson and Robert Galeta (Minneapolis: University of Minnesota Press, 2010), 189.
19 Gilles Deleuze, *Spinoza: Practical Philosophy*, trans. R. Hurley (San Francisco: City Lights Books, 1988), 90.
20 In an interview, Jia mentioned that it was Chen Kaige's *Yellow Earth* that inspired him to become a director. See Michael Berry, *Speaking in Images: Interview with Contemporary Chinese Filmmakers* (New York: Columbia University Press, 2005), 185.
21 Zhen Zhang, "The Negating Subject in Progressive Time: Jia Zhangke's *Xiao Wu*," *International Journal of Humanities and Social Science* 1, no. 18 (2006): 165.
22 On the distribution of global capitalism and the role of China, see George Ritzer, ed., *The Blackwell Companion to Globalization* (Malden, MA: Blackwell, 2007), 105.
23 Ibid.
24 Georg Simmel, *The Philosophy of Money*, 3rd ed., trans. Tom Bottomore and David Frisby (New York: Routledge, 2004), 449–50.
25 Ibid., 77.
26 Mary Ann Doane, *The Emergence of Cinematic Time* (Cambridge, MA: Harvard University Press, 2002).
27 Dorothea Olkowski, *Gilles Deleuze and the Ruin of Representation* (Berkeley: University of California Press, 1999), 21.
28 Ibid., 95.
29 Deleuze, *Cinema 2*, 82–83.
30 Zhen Zhang, "The Negating Subject in Progressive Time," 162.
31 Deleuze, *Cinema 2*, 77.
32 Ibid., 78.
33 Territorialization and reterritorialization are both terms used by Deleuze and Guattari to depict the function of capitalist logic, to make everything capitalized and consumable. See Gilles Deleuze and Félix Guattari, *A Thousand Plateaus: Capitalism and Schizophrenia*, trans. Brian Massumi (Minneapolis: University of Minnesota Press, 2004).

34 Frédéric Vandenberghe, "Deleuzian Capitalism," *Philosophy & Social Criticism* 34, no. 8 (2008): 879.

35 The concept "affect" widely used currently in academia can be categorized into two groups: the psychobiological tradition pioneered by Eve Kosofsky Sedgwick and Adam Frank, and the Spinozist-Deleuzian bodily ethological tradition restated by Brian Massumi and so forth, thus the consequent works are roughly generalized into eight groups by Melissa Gregg and Gregory J. Seigworth, eds., in "An Inventory of Shimmers," in *The Affect Theory Reader* (Durham, NC: Duke University Press, 2010), 5–8.

36 Bruce Baugh, "Body," in *The Deleuze Dictionary*, rev. ed., ed. Adrian Parr (Edinburgh: Edinburgh University Press, 2010), 35.

37 Felicity Colman, "Affect," in *The Deleuze Dictionary*, rev. ed., ed. Adrian Parr (Edinburgh: Edinburgh University Press, 2010), 11.

38 Brian Massumi, *Parables for the Virtual: Movement, Affect, Sensation* (Durham, NC: Duke University Press, 2002), 6.

39 Berry, "Jia Zhangke and the Temporality of Postsocialist Chinese Cinema," 117.

40 Smith, "VIFF 2012 Review."

41 De Luca, "Sensory Everyday," 158.

42 Michel Foucault, *Discipline and Punish: The Birth of the Prison*, trans. Alan Sheridan (New York: Vintage, 1995), 138.

43 Picard, "Programmer's Note."

44 Michel de Certeau, *The Practice of Everyday Life*, trans. Steven Rendall (Berkeley: University of California Press, 1984), 30.

45 Here, I concur with what Susan Ruddick says about affect politics. In her essay, Ruddick asserts that those "political projects emerge as counter-actualizations" and cannot be simply reduced to an actual regime in the current world. See Susan Ruddick, "The Politics of Affect: Spinoza in the Works of Negri and Deleuze," *Theory, Culture and Society* 27, no. 4 (2010): 21–45.

46 Dennis King Keenan, *On the Question of Sacrifice* (Bloomington: Indiana University Press, 2005), 1.

47 Ibid., 55.

48 Ruddick, "The Politics of Affect," 21.

49 Elizabeth Grosz, *Becomings: Exploration in Time, Memory and Futures* (Ithaca, NY: Cornell University Press, 1999), 11.

50 Keenan, *On the Question of Sacrifice*, 3.

CHAPTER 7

"Disposable" Bodies on Screen in Xu Xin's *Karamay*: Biopolitics, Affect, and Ritual in Chinese Central Asia
Darren Byler

Abstract

Based on an analysis of political speech and embodiment in the film *Karamay*, in this chapter I argue that ritualized ways-of-being, which rose to the fore in Maoist China, continue to form a deeply felt common affect for marginalized people despite rapid changes in the built environment and economic structures of mainstream Chinese society. In an effort to explore these claims, I analyze the way the monumental documentary film *Karamay* describes the long duration of a historical trauma, injustice, and alienation through its embodiment by a group of Han and ethnic minority oil workers and their families. I then consider the way this ritual embodiment relates to an affective atmosphere of failure for those on the margins of economic development and social justice in Chinese Central Asia. In order to parse the sources and forces of this shared experience, the chapter considers the valence of the biopolitical concept of "disposability" in tension with the anthropological concept of "ritual." It argues that a refrain that emerges from a close reading of embodiment in contemporary independent cinema in Reform-era China is an effect of political rituals that fail to provide the sense of well-being they promised in the Maoist past. Yet, despite their failure, intimate portrayals of the motion of these rituals still hail the viewer as an embodied *phronetic* struggle for existential stability.

Keywords: Affect, Ritual, Disposability, Biopolitics, New Documentary, Xu Xin, Xinjiang, China

Introduction

One of the dominant themes that has emerged in many recent neorealist films and documentaries in the Chinese-language independent cinema of the People's Republic is a focus on forgotten spaces and alienation in the midst of rapid economic development. Films from influential directors such as Jia Zhangke and Wu Wenguang have led the way in escaping the programmatic *telos* of both critical and socialist realism. Instead, these filmmakers have promoted an "amateur" (*yèyú*) or "on-the-spot" (*jìshí zhǔyì*) *phronesis* (the social knowledge and ability to act politically). In so doing they have developed a set of practices that privilege the immediacy of direct personal engagement over high production values, melodramatic storytelling, and neat resolutions that typified earlier forms of realist Chinese cinema (*xiànshí zhǔyì*).[1]

By focusing on the lived experience of rapid economic change these films provide a powerful assessment of the efficacy of modernization. Yet, in the analysis of these films, direct attention has rarely been paid to the "stickiness" of pre-Reform comportments that continue to intervene in the embodied rituals of those on the margins of this radical social change; instead, analysis of an "aesthetics of disappearance" and "transformation" as modes of cultural production have been a central focus. Drawing on the work of Paul Virilio and Gilles Deleuze, among others, many scholars have (often quite brilliantly) analyzed Chinese New Documentary and independent cinema in terms of an emerging Chinese urban aesthetics rather than the long duration of ritualized behavior.[2]

This chapter joins this discussion by arguing that ritualized ways-of-being, which rose to the fore in Maoist China, continue to form a common affect for marginalized people despite rapid changes in the built environment and economic structures of mainstream Chinese society. I explore the valence of the biopolitical concept of "disposability" in tension with the anthropological concept of "ritual" to argue that a refrain that emerges from a close reading of embodiment in contemporary independent cinema in Reform-era China is an effect of political rituals that fail to provide the sense of well-being they promised in the Maoist past. Yet despite their failure the intimate portrayal of these rituals in action still hails viewers in an embodied phronetic struggle for political and existential stability. In order to explore this claim I consider the way the monumental documentary film *Karamay* (2010) describes the long duration of a historical trauma through its embodiment by a group of Han and ethnic minority oil workers. I then consider the way this ritual embodiment relates to an affective atmosphere of failure for those on the margins of economic development and social justice in Chinese Central Asia.

Bodies on Screen in *Karamay*

Xu Xin's *Karamay*, is a meditation on the relationship humans have to the failure of ideology-driven Modernist political projects in our current historical moment. On December 8, 1994, the city of Karamay, the heart of the oil fields in Northwest China's Xinjiang Province, was the site of a horrific fire that killed 323 people, 288 of whom were schoolchildren. The carefully selected, high-achieving students present that day, clad in red and yellow, were performing dances from Mao's Eight Model Operas and singing Red Songs for state officials in a concert hall when a thin curtain positioned too closely to a 600-watt spotlight caught fire. As they moved in the synchrony of mass choreography, their red scarves tied neatly in place, acrid smoke from highly flammable insulation began to fill their lungs. Countering instinctual panic with Maoist discipline, the children in the audience were told to remain seated while the officials exited first. Due to lax safety standards, locked doors, and the delayed arrival of the fire department many of the children never escaped. When help finally arrived forty-five minutes later, the bodies of trampled and burned children were piled over a meter deep around locked metal exit gates; most died from smoke inhalation and the weight of bodies on top of bodies rather than the fire itself.

None of Karamay's city officials died in the fire. Despite initial admissions of guilt and promises of state-level martyr status—which carries with it economic and social security for the families of those who died—after the fire the story was heavily censored in the Chinese state media and street protests were met with brute force. Zhou Yongkong, head of PetroChina, the state-owned company that controlled post-Reform Karamay and today monopolizes China's oil, quickly stepped in.[3] Speaking on behalf of the children who died, Zhou thundered in archival footage featured in the film, "Those children are in heaven hoping for Karamay's stability." Following these remarks and the demotion of Karamay's mayor, mourning the loss was taken as subversive to the goals of state stability, and the parents who demanded justice were marked as deviants under the Reformist social contract. The moral responsibility for the tragedy had been made to fall largely on the truncated family networks of settlers and already under-privileged local minorities affected by the fire. The families of

Karamay were not allowed to publicly mourn their children, and instead were treated by local officials and other members of their work units as embarrassments and in some cases, as mentally deranged. The mayor's brother, who built the Friendship Hall and bribed the safety inspectors, was never formally charged.

With the exception of a minority of Uyghurs and Kazakhs, the majority of Chinese speakers in Northwest China come from elsewhere. Their families came to China's far Northwestern province of Xinjiang (*New Dominion*) in the 1960s to work in the oil fields and protect the Chinese frontier. Trading rural social networks for the future benefits of membership in the industrial proletariat, these parents placed their lives in the hands of the Party. They committed themselves to a national-communist project thousands of miles from their natal homes. They developed skills for coping with displacement. They disciplined their bodies and the bodies of their children as biopolitical weapons in a war with nature. Out on the frontier, rituals of patriotic citizenship took on an intensified significance; these pioneering settlers were on the front lines of the nation. If their sacrifice was not completely recognized in the central nodes of Chinese society, it was nevertheless deeply felt at its margins. Although times were extremely tough in Xinjiang, Maoist *biopolitics*—thought broadly as a system for managing the health and welfare of a population conceptualized as a social whole—enabled Han settlers to realize a poor, yet durable, existence. Xinjiang was not fraught with some of the insecurities that affected other parts of China during the Great Leap Forward and Cultural Revolution. Relative to Beijing, Xinjiang was a stable place for settlers (less so for indigenous minorities under the new regime); rituals of sacrifice and interdependence that came with the infrastructure of rationing and cooperative social organization were largely effective in maintaining a sense of well-being.

Yet as the vitality of the Maoist social-national project dissipated in Xinjiang during the Dengist Reform era, some of these same people found themselves superfluous, caught up in forces much larger than themselves and what they were promised. After the 288 children of the city of Karamay[4] died in the horrific fire in 1994, unaffected officials and citizens moved on with economic redevelopment, apathetic toward the lingering economic, social, and institutional inequalities that continued to affect the families of the dead. In its late-Socialist iteration, the *ethos* of their work unit, PetroChina, and its subsidiary support units no longer seems to account for their well-being. The parents Xu Xin interviewed in this film feel stuck, unable to move with flows of power and wealth that buoyed the futures of so many

"Disposable" Bodies on Screen in Xu Xin's *Karamay*

Reform-era Chinese. More than an exposé of the tragedy of loved ones lost, this film is about the corporeal embodiment of social abandonment and failure.

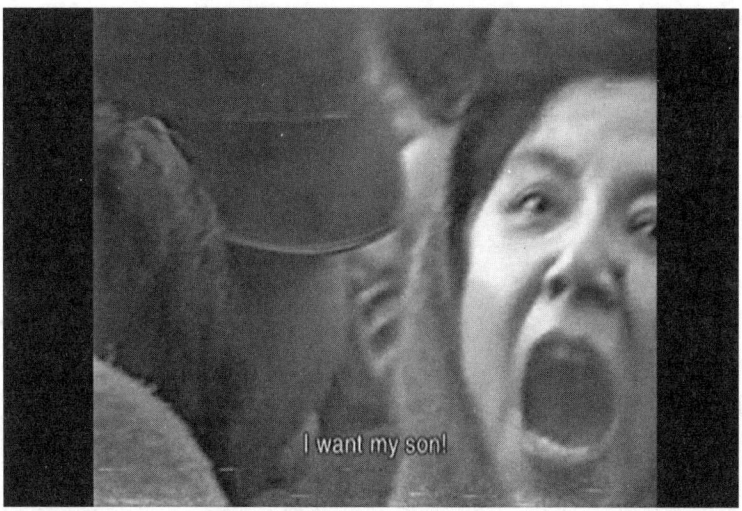

Figure 7.1. One of the Uyghur parents screams in Mandarin in a street protest shortly after the fire in Karamay in 1994—86 of the 288 children killed were minority children in Xu Xin's *Karamay*. Image courtesy of dGenerate Films/Icarus Films.

Karamay lays bare the margin of raw violence of human interaction that accompanies the disenfranchisement of collective ideals (Fig. 7.1). In this late-humanist moment "a concern for human beings finding themselves and becoming free in their humanity" is becoming increasingly untenable for those on the margins.[5] The precariousness of those without positive social ranking in China's late-Socialist context—the "common people" (*lǎobǎixìng*) as these parents self-identify—is becoming more acute. What *Karamay* does, then, is point our attention to what "disposable people"—to use Rey Chow's turn of phrase—look like in Chinese film.

In developing her concept of "disposability" Chow argues along with Étienne Balibar and Bertrand Ogilvie that the human condition of our present moment of global capitalism is one in which the lives of "millions of human beings *are superfluous*."[6] Reading Balibar and Ogilvie's claims through the lens of Martin Heidegger's phenomenology, Chow argues that in our current moment of global capitalism humans are increasingly entering into a state of existential "homelessness." That is, the *being-in-the-world* of humans is increasingly rendered in a *state of "oblivion"*;[7] a state in which the *techne* and *poiesis* of political action, social organization, and human cultural processes are muted and ineffective.[8]

Discussing the way this phenomena is manifested in Chinese late-Socialism through an analysis of the cruel life world of Northern Chinese coal miners in *Blind Shaft*, Li Yang's neorealist 2003 film, Chow argues that "the major culprits here are the structural deficiencies that pervade the entire industrial production system in China."[9] As in other developing countries, a dominant feeling and experience among many "disposable" people in contemporary China is that the population exceeds the capacity of institutions to provide social welfare or biopolitical health. Yet as Chow points out, the implications of films like *Blind Shaft*, and I would add, *Karamay*, should not be thought of as particular to "third-world" states-of-exception. Instead, she argues, what we are seeing on film is "a dramatization … of the predicament of human community formation in general."[10] The embodied situations of both *Blind Shaft* and *Karamay* are entangled in the excavation of energy through which industrial, commercial, and cultural development are made possible. It was, after all, the Modernist project of securing oil and gas as resources for the nation that brought the people of *Karamay* to China's Northwest. In the end, the city of Karamay, like countless locations across the planet, is an industrial boomtown inextricably linked to political-economic development.

Certainly, throughout this process of development the people of *Karamay* have experienced sacrifice for their nation and their families. Yet, as Chow argues, it is only in our contemporary moment of the ascendance of global capitalism that marginalized people around the world have witnessed "the very mutation of the concept of 'human' … as the unconcealed process of species differentiation that is happening at the rupturing between … humanity-as-progress, or hope … and the ubiquitous biopolitical warfare around natural and other resources and, above all, around kinship and other types of group survival."[11] For the parents in Xu Xin's *Karamay* it is these basic intersubjective, embodied social relations that are at stake. As these relations are threatened by the welfare abandonment of the social state and its institutions, parents find themselves attempting to reclaim an attachment to the "kinship family," which throughout Chinese history has been thought of as an "inviolable basic social unit,"[12] and the corporeal rituals and gestures that give this affective attachment its embodiment.

Although the rituals that support the "right-to-a-family" and by extension "the good life" have undergone numerous involutions and deviations over the centuries, they have nevertheless been central and relatively stable modes of reproducing the relations of the individual to the state and of the individual to the family. These rituals of speaking and saving are what are embodied—a process of incorporating the social and material world corporeally—in *Karamay*. If "speaking bitterness" and "saving face" were actions that brought dominant cultural tropes into the historical lived experience of Chinese subjects, what do these rituals look like in this historical moment of capitalist expansion and the reterritorialized space of social welfare erasure? How are they embodied by people who have sacrificed so much, and, in Heidegger's sense of *being-in-oblivion*, seem to be so far from home?

Given its focus on the long duration of processes of failure, perhaps it should come as no surprise that *Karamay* is a difficult film to watch. It took me over a month to get through all five hours and fifty-six minutes. Why does it feel this way: compelling and repellent, tedious and captivating? Speaking about his feelings making the film, Xu Xin said: "I don't know how to understand happiness. Although the content of the film is very painful, I had a joyful feeling while making the film itself. I don't know what to think of this."[13] Such ambivalence suggests that the film is more than a monument to tragedy. Though the topic is unsettling, the *pathos* that comes out in the slow minutes of the film is so visceral viewers find it hard to look away. I suggest that this training of attention is drawn from the points in which *affect*—as a range of feeling—comes to the surface and sorts itself out in emotion and then resubmerges as an unspeakable current in the nervous system. Engaging discussions of affect and ritual, I describe the way affect appears in human embodiment and corporeal sacrifice as a "wisdom of the body." Following this overview of my terms of discourse, I then turn to the specifics of *Karamay* as a particular embodiment of trauma and ritual therapy in Sinophone film. I conclude by arguing that the mirroring of the affective atmosphere of the production process which can be observed through the viewing process is important for understanding "disposable bodies" on screen in *Karamay* and Chinese independent documentary film more generally. Despite the particularity of the historical situation of *Karamay*, the embodied experience of viewing the film invites an intimate knowledge, an affective atmosphere, of the embodied, corporeal life from which no "exotic other" can be parochialized. Put simply, on the level of the body, viewers are invited to relate with the viewed.

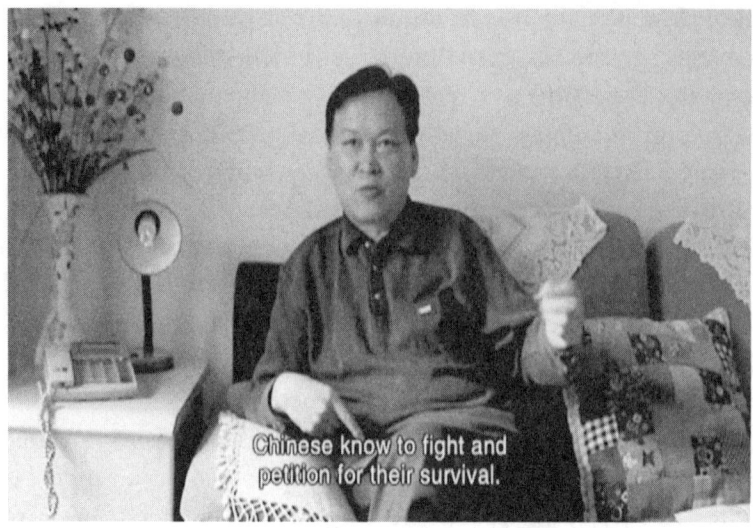

Figure 7.2. One of the most outspoken parents discusses the historical legacy of protest in China while shaking his fist in Xu Xin's Karamay. Image courtesy of dGenerate Films/Icarus Films.

Affect and Ritual Embodiment

The anthropologist Hugh Raffles has noted that "people enter into relationships among themselves and with nature through embodied practice. … it is through these relationships that they come to know nature and each other."[14] These relationships, knowledge, and practice are always mediated "not only by power and discourse, but by affect … the perpetual mediator of rationality."[15] Defining this "affective sociality" as "intimacy," Raffles describes the ideology inhabited by localized rituals as "always within a field of power…always in place…always embodied… always, above all else, relational."[16] Raffles argues that if "relationality is a social fact," then "there is no universal against which intimacy is parochialised."[17]

Moving toward a more precise description of the relationship of the affective to fields of power, the anthropologist William Mazzarella tells us, *affect* is neither completely external to mediation nor simply a discursive *effect*. Reviewing ethnographic writing on the subject from Émile Durkheim, Mazzarella concludes there is a "nonsubjective sensuous mimetic" power to this register of the social; particularly as it is converted to *ritual*.[18] He writes, "the language of ritual is the language of power;"[19] it is an untimely grammar that works through the mediation of the body to exert power in the world. As anthropologists have long observed, rituals—broadly defined as actions intended to reproduce social norms and political conventions—are what organize and animate a society in the absence of an intervening ideology. The range of action and feeling we see arising out of the socio-political atmospheres of late-Socialist Xinjiang are therefore a local iteration of historical forces and contemporary circumstances. Mainstream values such as social stability and economic development are meeting a ritualized system that is no longer amplified. The microphone that projected messages of bitter Socialist struggle has been unplugged, yet the embodied expression of these performative rituals still remains at the margins of Chinese society: confronted with a public space of petition in front of Xu Xin's camera, parents are first animated by the ritual of baring their scars only, in turn, to sag defeated as the ritual fails (Fig. 7.2).

If the institutionalized practice of "telling bitterness"[20] (*sùkǔ*) was a form of performative Socialist ritual, a mimesis or imitation of the affective that fitted power into place, then it seems likely that affective feelings entwined in Socialist subject-making are not something that have been completely jettisoned by Chinese late-Socialist reforms. Robert Chi's reading of *Red Detachment of Women*—Jin Xie's 1961 Maoist film—compels us to acknowledge that official narratives ascribed to "history" cannot be detached from mnemonics. Noting the way these performances of showing and telling bitterness serve "to focus particular attention on the body as the site of both memory (as suffering, as an effort against negation) and sociality ... the mass public experience," Chi reads the legacy of Socialist ritual aesthetics as containing both a catalyst for "somatic gesture and as emotional stimulation."[21] If one of the dominant visual-somatic ritual elements of Socialist China was "the baring of scars and the shedding of tears,"[22] then the parents bearing witness in *Karamay* must be considered as disjointed, yet derivative, of that same Chinese "spectatorial body." As the disciplinary power of this past discourse dissipates and joins with the discipline of the Neoliberal state to come, we see Chinese citizens turning to the discipline of new ritual forms of mediation. These are forms that involve the *techne* and *poesies* of the digital video camera, the video archive, the *presence* of the interviewer-as-interlocutor; yet, as Xu Xin's film shows us, the memories of those earlier forms of petition and protest have not yet been completely erased. When a man who is speaking bitterly hurls the Chinese sign for "six" at viewers with an up-turned hand, his pinky and thumb extended as he corporeally emphasizes the fact that fire fighters were stationed only five or six minutes away but still took forty-five minutes to arrive; when parents "bare their scars" through the onomatopoeic invocation of their phantom children running down the stairs (*dùng, dùng, dùng*) for the last time and then turn inward, heads bent, their elbows on their knees; when they hold each other heaving in pain and scream that "heaven is blind;" when they explain that their households are "broken"—their hands moving out from their chests in open-handed gestures; they are showing us that a shared affective experience, crystalized in the rituals of a Maoist political body, resists easy erasure. They are showing us that people always attempt to stay attached to the conventionality of life even when that form of life is mutating. They are showing us that the ordinariness of the long duration of social crisis forces people to struggle for existence using obsolete forms of composure even as a tractable future is steadily contracted.

Starting from the assumption of the theorist Lauren Berlant that "affective atmospheres are *shared*, not solitary, and that bodies are continuously busy judging their environments and responding to the atmospheres in which they find themselves,"[23] we can see that the ubiquitous horror of the fire provided a common historical grounding across class and ethnic divides for the families of those who died. At the same time, the easy disposal of children's bodies, the absence of death certificates for those who died, the quick dismissal of parents' claims to justice, the shunting to the asylum of those crazy with grief, and the way their bodies are wracked by nervous and psychic maladies, tells us something also about the contemporary mood and mode in which "common people" experience their value as Chinese citizens, people, and loved ones. Although it may be tempting to consider the collectively experienced disaster and subsequent position of the parents and children of *Karamay* as an exception to the Chinese narrative of progress building on a deep body of literature from Ann Anagnost, Pun Ngai, and many others, the symptomatic experience of alienation and impasse felt by the parents of *Karamay* must be considered in light of the erasure of the Maoist class structure and the abandonment of social welfare concerns. This political-economic structural bifurcation geared toward rapid development and increased individual-family network responsibility is felt ubiquitously in contemporary Chinese society. Framed in this way, the situation in *Karamay* can be read as just one acute iteration of the simultaneous reshaping and durability of Chinese conventions.

"Disposable" Bodies on Screen in Xu Xin's *Karamay*

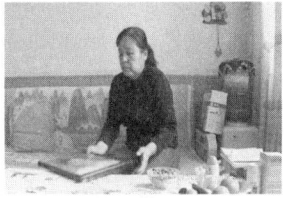
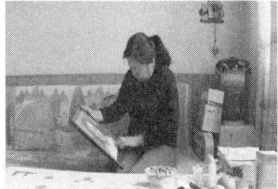
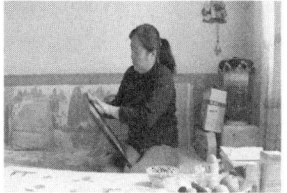

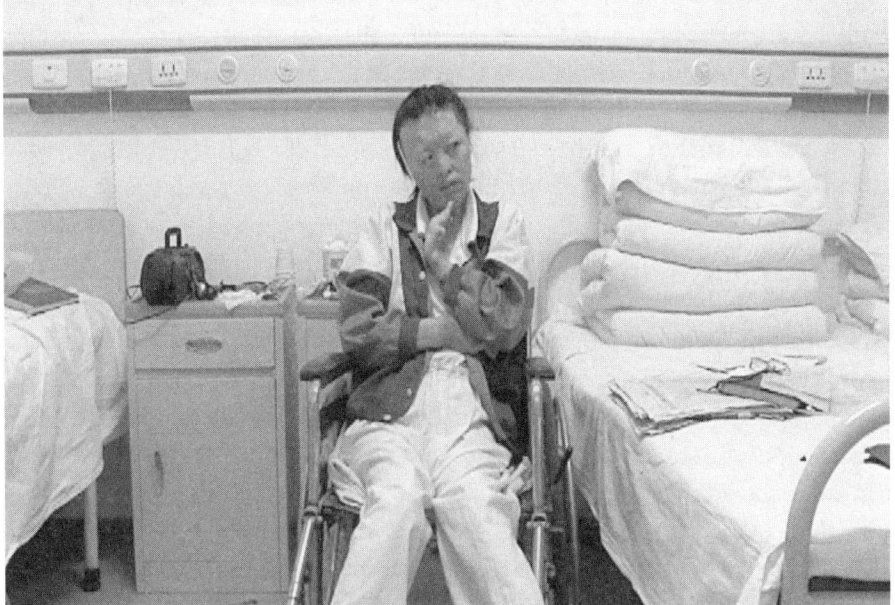

Figures 7.3-4. Subjects seem to forget about the camera in moments of nervous distraction and unspoken melancholy in Xu Xin's Karamay. Images courtesy of dGenerate Films/Icarus Films.

Ritual Therapy under the Affect of the Impasse

In Bérénice Reynaud's 2010 reading of Chinese independent cinema as typified in Wu Wenguang's approach to documentary craft, we see that the aggressive silence of the filmmaker-as-therapist can be a quiet intervention that allows the subjectivity of the observed to emerge. Rather than invoking a scripted reaction, the therapy of silence can allow a subject "to express a discourse of desire."[24] Citing a pivotal scene in Wu's *Fuck Cinema* where the main protagonist confronts Wu by speaking directly into the camera, Reynaud argues with Lacan that it is in this encounter, where discursive desire falters and the imaginary fails to surface, that a version of "the Real" can be glimpsed. What makes *Karamay* distinct from *Fuck Cinema* is the repetition of the therapeutic silence necessary for a ritual circuit—as a repeated set of bodily techniques—to be performed on film in a wide range of similar yet slightly different circumstances. While *Fuck Cinema* is largely framed around a single decontextualized individual and aggressively questioned migrant women, *Karamay* is centered by a shared duration of a collective experience of trauma. This difference, along with the temporal scale of the film, are what make *Karamay* a limit case for analyzing the embodiment of ritual in Chinese New Documentary cinema.

By rendering the iterative collective process of *disposability* visible in what Gilles Deleuze refers to as repeated filmic "time-images," we can begin to identify a movement of affect as a range of feeling between anger and failure before and after it comes to be recognized as either of these discursive emotions.[25] It is in these moments when the play of ritual runs its course that an embodied gesture that resists symbolization appears: a movement of a hand, a turning away, a drawing into the body, a flash of life void of being, the sag of failure. Seeing the repeated circuit of this embodied turning from norms of resolution and social integration conveys something of the trajectory of existence for these parents; seeing the repetition of these somatic gestures captured in a time-image on film (rather than described in a text) conveys something of the immediacy of this sensorium. To my thinking, these instances are analogous to Barthes's idea of the *punctum* in a still image: the kernel of "the Real" or, on a discursive level, "the reality effect," which survives mechanical automatism.[26] It is a kind of animacy that sears something into viewers' brains; it triggers empathy and intercorporeality. Wrapped around the failure of "telling bitterness" in *Karamay* it evokes an affective tuning that is more than the sum of its parts. These rituals of anger that inevitably turn to failure communicate the corporeal feeling of bodies rendered *disposable*. As the feeling of the *oblivion-of-being*, of being without place, is invoked, viewers are invited to share, to relate to, the embodied pain of the impasse.

The scenes of embodied disposability that rise to the surface in *Karamay* (Fig. 7.3-4), that bring forward the shattered affect of those who passed through the "door to hell" (as they refer to the lowered gate on the Friendship Hall), are the poignant images such as that of a mother, who after speaking for many minutes, lapses into silence and forgetting about the camera compulsively strokes at phantom dust on the frame of a photo of her dead daughter—an image of the void of being-without-language; there is the image of a father leaning back his eyes pinched in frustration, then defeat—an image of the slump of powerlessness; there is the image of a young woman who survived the fire who lapses into melancholy, thinking about her object of desire: Nanjing University and the promises of the good life she will never have. Her face, masked by ruined and grafted skin, contained by the anonymity of her secluded hospital room, still conveys an image of the pathos of human longing for a barred object of desire.

Xu Xin approached each interview with the same gray-scale palette, straight-ahead composition, and minimal direction. Like Huang Weikai and many other contemporary Chinese documentary filmmakers, Xu Xin's educational background was in painting and the fine arts. It is perhaps because of this training in color and frame that *Karamay* is so effective in constructing nested worlds of color and gray scale. With the exception of four scenes of Karamay's cityscape and flashes of horror from the parents' personal video archive of the fire, the world of the film is muted gray walls—there is no horizon for these parents. They are alone, stuck with their grief, and outside the affluence and forgetfulness afforded by Karamay's vast oil wealth. The repetition of framing and the minimalism of their colorless world have the effect of amplifying the tension in the non-linear narrative of the film. That is, the tension of the narrative in this *diegesis* displays not only the textual and ideological position of the film, but also the parameters that direct viewers as subjects and expose the embodied presence of the filmed. As parents repeatedly smack the backs of their right hands into the open palms of their left and describe the injustice of their state, as they sketch the contours of the building where their children died, "the pull" (*lā*) of ripping off the locked gates that trapped their children, viewers sense that *Karamay* is the same brutal story told over and over again, yet slightly refracted by the many angles of singular storytellers.

In order to foreground the long-duration of trauma, Berlant has described such experiences of precariousness as "an impasse" (Fig. 7.5). This refocusing away from crisis-events, such as a fire or some other drastic action that seems to have a clear cause and effect, toward the long aftermath where interrupted norms of life are reconfigured, trains viewers' eyes toward inexplicable moments that appear outside narrative genre. The way Xu Xin captures dramatic gestures of anger, fingers pointing and fists clenched, followed by quiet gestures of failure, of heads buried in hands, of eyes looking to the side, lost in the middle distance, shows us how people are struggling to adapt to the impassivity of what we see as the Real. In Berlant's words, "An impasse is a holding station that doesn't hold securely but opens out into anxiety, that dogpaddling around a space whose contours remain obscure."[27] As the parents and children of *Karamay* come to terms with the ineffectiveness of old modes of ritual protest, the impassivity of the new situation simultaneously demands action and delay.

After a social catastrophe there is always a period of adjustment. This is the shared affective atmosphere in which we see the figures of *Karamay* forced into new gestures of composure, new forms of *phronesis*: for example, we see parents describing the way a person's shoulders are pulled back when they are manhandled by police during a protest; we see the embodied mimicry of suicides attempted. But even more affecting is the heavy gaze of *the disposed* at the end of the ritual circuit. There is a numb lifelessness in many of those looks—their eyes are open but they are not looking at anything. It is this diegetic world that exposes viewers to moments of affective rupture as autonomous time-images. As rituals of "speaking bitterness" are shown to be ineffective, the unarticulable affect of failed attachments rises as a punctuation that transforms viewers from passive spectators to active witnesses of powerful forces at work in the time-space of these disposable bodies on screen. As the time of repetition (*chronos*) is interrupted by these small temporal events (*aion*), viewers are presented with a disorienting vision of the present within the duration of the lives we see on film.

As Xu Xin thinks of it, what he was trying to establish through this approach to interview was a "spiritual connection" that comes from honest exchange and direct recognition. He writes,

> I came to feel the kind of emotion that [the parents] had really deeply. Before every visit [I told them] "you don't have to speak to me." I told them …very explicitly, I'm making a documentary about the fire in Karamay. They were all very clear about what I had in mind. Because of this when I was shooting, they all looked straight into the camera. This aspect was extremely important: I looked straight into their eyes, we really had a spiritual connection, I used my soul to listen to their stories. If, say, we didn't have this exchange, they also could have spoken, but speaking is not the same as communicating. The things they kept in their hearts for the past ten years all of the sudden burst out.[28]

In order to get at this trust and catalyze this "bursting out" of a discourse of desire, Xu Xin positioned himself directly behind his camera and "just sat" there. The "soul tending" that Xu Xin is respecting here is the same feeling that invites viewers as they move deeper into the world of the film. As viewers learn to be intimate with failure, the disposable bodies on screen begin relating or "connecting" with viewers on a corporeal level. By allowing the camera to linger on center-posted parents during small moments of not speaking, framed by white walls on broad Chinese couches, Xu Xin allows us to become proximate to the feelings of these people.

Clearly there was much that these parents wanted to say; rituals are on one level an iterative public performance (in this case, for the camera). Yet, the readiness-to-hand of their rhetoric of "speaking bitterness," the naturalness with which they "bared their scars," and the sag of defeat that comes through as they lapse into silence point more deeply to an embodied intercorporeal experience rather than to narratives that operate solely on the ideological-political register of identity performance. Of course the embodied cannot be detached explicitly from the ideological.[29] Yet what this film points us toward is an understanding of the weight of embodied experience that is immanent in the performance of political identity—be it national, ethnic, socialist, or capitalist. These parents—Chinese subjects officially recognized as Han, Hui, Kazakh, and Uyghur citizens—felt as though they had nothing to lose by speaking to Xu Xin, and they felt a catharsis in his public recognition of their personal stories. These rituals of petition, largely emptied of political force in this late-Socialist moment, still convey a catalytic pathway for the "somatic gesture and as emotional stimulation" of cultural replication.[30] Yet on the threshold of an economy of "development" and "progress" the repetition of these ritual performances of "baring scars and shedding tears" are now marking these parents as *disposed* rather than heroic.

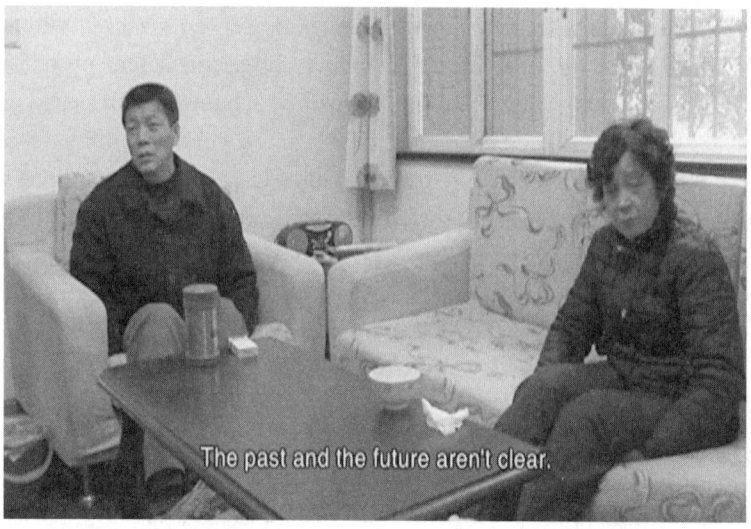

Figure 7.5. The impasse in Xu Xin's Karamay. Image courtesy of dGenerate Films/Icarus Films.

Conclusion

I began this chapter with a discussion of *Karamay* as a film about the way bodies are caught in political projects and how blockages in these forceful systems can result in anomie and affective inertia. Yet *Karamay* is also a film about the relationship between cinema and witnessing. As Shoshana Felman writes about a similar project, Claude Lanzmann's *Shoah*, a film can embody "the capacity of art not simply to witness, but to take the witness's stand: the film accepts responsibility for its times by enacting the significance of our era as an *age of testimony*, in which witnessing itself has undergone a major trauma."[31] In the context of precariousness, witnessing the objective reality of people under trauma becomes "in all senses of the word, a *critical* activity."[32] Yet like Laura Marks, I would emphasize a bit more strongly than Felman that both *Shoah* and *Karamay* do more than "authenticate," in a legal sense, the truth of the trauma survived by those filmed. Rather, these films unfold a "sheet of past from a peak of present."[33] The ethical nature of these films exists therefore "not in authenticating testimonies, but rather in demonstrating that some events are too terrible to be fully actualized…while insisting that they must be conceived of."[34]

In *Karamay* people find themselves in states of affective inertia—a nervous abnormality that renders them unable to act or react to their social world in normalized ways. As the bodies of their children and, likewise, their kinship family more generally, are rendered disposable, the parents in *Karamay* reel between anger and defeat. Although these feelings are certainly not evenly distributed across time-space among "common people" in Northwest China (many have found ways to detach from Maoist comportments and "move on" in the new political economy), in the sample population of the film we see a qualitative, palpable presence of these cycles: first fingers pointing, fists clenched, palms smacking, then heads buried in hands and, finally, almost universally, a vacant gaze disengaged both with the world of the film and, viewers are invited to extrapolate, the broader social world. What we see is that the worlds of the parents in *Karamay* are punctuated by states of psychic rupture and stall. In Xu Xin's gray-scale long-takes, time-space is therefore seen as charged with muted affective intensities and subtle feeling that appear to have an anomalous, unmotivated, autonomous temporality.

The long duration of the viewing experience of the durative present is important for understanding disposable bodies on film in *Karamay* and Chinese independent documentary film more generally. Like much of early Chinese independent film from Jia Zhangke to Wu Wenguang, Xu Xin's approach to documentary craft lends itself to a detachment of a humanist ethos of filmmaker-as-intervening benefactor. Instead, like the vast majority of Chinese independent cinema, it attempts to make explicit the terrain of the sensible in Chinese late-Socialism without foreclosing its message with programmatic narrative or ideological indictment. It is instead a documentary that bears witness to the duration of lived experience in Northwest China and allows critique to emerge from the material, embodied world on film.

When asked about the reasons he made the film, what sort of contribution he and his informants were hoping to achieve, Xu Xin declined to comment. He said that beyond telling their stories, those "deeper reasons I'm not capable of analyzing. Furthermore I don't want to analyze those reasons." Rather than pointing us toward a hopeful future or even a recognized clarification of the past, the film instead directs viewers to the embodied particularity of this shared historical moment after the capitalist mutation of the social state in a discrete social location in contemporary China. Even more important, it invites viewers to share in their feeling of first the animas of anger and then the slackness of anomie as the contours of a gutted existence come into view.

The parents' repeated demands for martyr status or at the very least, death certificates, for their children, are claims that operate on both a utilitarian and spiritual register. While the first claim is toward a project identity organized by a collective attempt to claw out a space of social security, the second is an intercorporeal operation that implicates all citizens of social states. By requesting martyrdom the parents are demanding that their children's sacrifice be recognized by the sovereign state as a contribution to the spiritual mission of the nation. By first promising then denying this recognition, the state is delineating its values in biopolitical management. For the state there seems to no longer be a utopian future toward which common citizens can sacrifice themselves. The new political economy depends less on ontological security and more on productivity. As Rey Chow puts it, "the future is contingent on the status quo … the continued solicitation, exploitation, and extermination of 'foreign' bodies that are considered as excess and disposable once they have served their utilitarian purpose."[35] For these parents, and viewers who share this embodied experience, the concept of the human itself is under mutation. Although manifested differently in other situations where alternate failed rituals operate in other margins, the intensity of feeling embodied in *Karamay* is indicative of an ordinary atmosphere of *disposal* in the shadows of the global capitalist iteration of biopolitical success.

Writing in response to Xu Xin's film and the way certain aspects of the Chinese world frequently disappear from view, Ni Ba had this to say about the film,

> When I asked a few of my friends if the name Karamay made any impression on them, a few friends said, "Is Karamay a country?" One friend said Karamay is a desert, still another said Karamay is a person's name. When I then told them about the great fire in 1994, they said: "Oh yeah!" They said that now they could recall it, vaguely…. Perhaps this is precisely our society's present condition. Development as the imperative, stability as the paramount priority, and patriotism that parrots the greatness of the motherland, have all served as powerful ideals for the duration of our great journey to revival. People with or without intention have chosen to forget about the way the lives of others confront the real and the future. Yet hearing the immediacy with which many regular people in Karamay know the way their children were blocked in the course of the events, and knowing that according to the official rhetoric better times should have come to that

place, I can't help but wonder what kind of success is bound up in [our economic development].³⁶

Cinematic witnessing, which we see exemplified repeatedly by Chinese independent films such as *Karamay*, is a presentation of objective reality that demands that the stratified order of things not return to normal. It makes us recognize that the normal discourses of the dominant are incomplete and inaccurate.³⁷ This radical cinema of witnessing is concerned with making visible "the sensible" as a terrain of *what can be shown and felt* on ethical, representational, and aesthetic registers. It presents a social project defined as a struggle for recognition and legitimation in which the "excluded part" of social systems demands a space of common relation. This sort of therapeutic intervention must be understood as a perpetual phronetic practice of sharing a feeling, sharing a cadence of a particular experience of the Real.

Many New Documentary films such as Zhao Liang's *Petition*, Wu Wenguang's *Fuck Cinema*, and neorealist fiction films such as Jia Zhangke's *The World* and *Still Life* use a long view of alienation and displacement to present an implicit critique of the disposability that accompanies rapid economic change. *Karamay* extends and amplifies these feelings of anomie by drawing out the ritual circuit of failure in a prismatic repetition of framing and narrative variation. By centering the film on the collective repetition of embodied rituals rather than the singular movements of isolated individuals, *Karamay* brings forward the way old feelings of collective affective atmospheres continue to operate in the durative present of contemporary Chinese traumas. Rather than describing an aesthetics *of disappearance* and *transformation*, *Karamay* hails viewers with an aesthetics *of feelings that remain*. The critical ethics of *Karamay* is one that belies the perception that problems experienced by disposed people are felt in largely singular, unmediated ways. Rather, it is in order to undermine the rhetoric of "free market" success and embody the stubborn shadows in narratives of progress, that Xu Xin gives us these "disposable" bodies on screen.

Notes

1 Chris Berry and Lisa Rofel, introduction to *The New Chinese Documentary Film Movement: For the Public Record* (Hong Kong: Hong Kong University Press, 2010), 3–13.
2 Here I am thinking in particular about work on contemporary cinema catalyzed by Ackbar Abbas and Zhang Zhen; Ackbar M. Abbas, *Hong Kong: Culture and the Politics of Disappearance*, vol. 2 (Minneapolis: University of Minnesota Press, 1997); Zhang Zhen, ed., *The Urban Generation: Chinese Cinema and Society at the Turn of the Twenty-First Century* (Durham, NC: Duke University Press, 2007).
3 Zhou was also head of national security and Xinjiang security in particular under Hu Jintao's administration—making him the ninth most powerful man in China's Politburo. In August of 2013 a graft investigation of his political-economics was initiated by the Xi Jinpin administration.
4 Karamay (Ch: *Kèlāmǎyī*; En: *Black Oil*) is historically a Mongol, Kazakh, and Uyghur city in what is today Northern Xinjiang. Due to its vast oil resources, over the past few decades it has become a seventy-five percent Han city dominated by state-owned enterprises. According to a 2011 list of richest cities, Karamay is now one of the wealthiest per-capita cities in China (Si Han, "China's Richest 20 Cities," 2012, http://business.sohu.com/20120327/n339020485.shtml.).
5 Rey Chow, *Sentimental Fabulations, Contemporary Chinese Films: Attachment in the Age of Global Visibility* (New York: Columbia University Press, 2007), 178.
6 Étienne Balibar quoted in Chow, *Sentimental Fabulations*, 167.
7 Ibid., 168.
8 Ibid., 173.
9 Ibid., 170.
10 Ibid., 173.
11 Ibid., 178.
12 Ibid., 174.

13 All direct quotes from Xu Xin in this chapter are my own translation drawn from a Chinese language interview conducted with Ni Ba, "'Kelamayi' daoyan su sin fangtan," Fanhall.com (2010), http://gsz2006.i.sohu.com/blog/view/154025600.htm.
14 Hugh Raffles, "Intimate Knowledge," *International Social Science Journal* 54, no. 3 (2002), 326.
15 Ibid.
16 Ibid.
17 Ibid., 332.
18 William Mazzarella, "Affect: What Is It Good For?" *Enchantments of Modernity: Empire, Nation, Globalization* (London: Routledge, 2009), 298.
19 Ibid.
20 "Telling bitterness" finds its most visible institutional iteration in the Communist People's Courts of the 1940–1950s. These village meetings were a stage at which "common people" would articulate the injustices committed by their landlords. One could argue, however, that this mode of gaining a sense of agency and recognition has much deeper roots in Chinese literary traditions and popular culture. Since Guan Hanqing's Yuan Dynasty "Injustice to Dou E," Chinese popular culture has featured instances of misrecognized subalterns speaking "truth to power" as dominant themes in staged performances. See Ann Anagnost, *National Past Times: Narrative, Representation and Power in Modern China* (Durham, NC: Duke University Press, 1997), 28–35, for a thoughtful account of how this ritual came to be embodied in revolutionary China.
21 Robert Chi, "*The Red Detachment of Women*: Resenting, Regendering, Remembering," In *Chinese Films in Focus II*, ed. Chris Berry (London: British Film Institute, 2003), 154, 158.
22 Ibid., 154.
23 Lauren G. Berlant, *Cruel Optimism* (Durham, NC: Duke University Press, 2011), 15.
24 Bérénice Reynaud, "Translating the Unspeakable: On-Screen and Off-Screen Voices in Wu Wenguang's Documentary Work," in *The New Chinese Documentary Film Movement: For the Public Record*, ed. Chris Berry, Lu Xinyu, and Lisa Rofel (Hong Kong: Hong Kong University Press, 2010), 175.
25 Gilles Deleuze, *Cinema 2: The Time-Image* (Minneapolis: University of Minnesota Press, 2010).

26 Roland Barthes, *Camera Lucida: Reflections on Photography* (New York: Macmillan, 1981).
27 Berlant, *Cruel Optimism*, 199.
28 Xu Xin in Ni Ba, "Interview with Xu Xin." Emphasis added.
29 If, as Louis Althusser described, "ideology" is in many ways the relationship of our empirical experience to "the Real," social training in emotional expression are crucial nodes in the development of our mimetic practices. Rituals taught through social interaction introduce subjects to a dense play of signifiers. Yet the immediacy of embodied affect at the bounds of imagination and desire still seems to exceed this symbolization. Louis Althusser, Lenin and Philosophy (London: New Left Books, 1971).
30 Chi, "*The Red Detachment of Women*," 158.
31 Shoshana Felman, "Film as Witness: Claude Lanzmann's *Shoah*," In *Holocaust Remembrance: The Shapes of Memory*, ed. Geoffrey Hartman (Oxford: Blackwell, 1994), 91.
32 Ibid., 92.
33 Laura Marks, "Signs of the Time: Deleuze, Peirce and the Documentary Image," in *The Brain Is the Screen: Gilles Deleuze's Cinematic Philosophy*, ed. Gregory Flaxman (Minneapolis: University of Minnesota Press, 2000), 205.
34 Ibid.
35 Chow, *Sentimental Fabulations*, 175.
36 Ni Ba, "Interview with Xu Xin." Emphasis added.
37 Slavoj Žižek in Jacques Rancière, *The Politics of Aesthetics* (New York: Continuum, 2006), 70–71.

CHAPTER 8

The Art of Eating in Malaysian Cinema: The Malaysian Sinophone Hunger for a National Identity
Lee Yuen Beng

Abstract

Over the years, Malaysian cinema has solely focused on the problems and lifestyles of the Malays, catered to the Malays, and featured Malay casts speaking in Malay. In effect, Malaysian cinema is a Malay-centric industry that has negated the cinematic presence of other Malaysian ethnic communities. This ethnocentricity is currently being contested by a select group of prominent Chinese Malaysian filmmakers. Their works have collectively led to the rise of the Malaysian Sinophone cinema through their shared desire for the articulation and recognition of "Chinese-ness" in Malaysian cinema. In reversing this cinematic negation, they employ transnational methods to narrate stories of Chinese Malaysians through previously unheard of Sinophone films that have similar narratives, themes, subject matter, and languages. This collective usage of "Chinese-ness" goes beyond its cinematic representation by directly contesting and renegotiating the Malaysian national identity while criticizing the political economy of Malaysia and its cinema. This contestation is carried out through scenes of eating, which are in essence representations about the need to satisfy a hunger—a desire and longing for equal identification and nationhood as Malaysians no longer marginalized by race- and ethnicity-based policies.

Key Words: Malaysian Cinema, Sinophone, Eating, Desire, National Identity

Introduction

Since 2000, a select group of Chinese Malaysian filmmakers have been directly challenging ethnocentricity in Malaysian mainstream cinema. Their desire for the articulation and recognition of "Chinese-ness" in Malaysian cinema through films such as *The Beautiful Washing Machine* (2004), *Flower in the Pocket* (2007), and *Ice Kacang Puppy Love* (2010) have been deconstructing racial and ethnic stereotypes and have led to the rise of a Malaysian Sinophone cinema. Made using transnational methods, this Malaysian Sinophone cinema has reversed the negation of the cinematic presence of Chinese Malaysians. The telling of stories about Chinese Malaysians, with narratives and characters from the ethnic Chinese community speaking Chinese dialects, also directly contests and renegotiates Malaysian national identity. The films of Ho Yuhang, James Lee, Tan Chui Mui, Ng Tian Hann, and Khoo Eng Yow—which are similar in terms of narratives, themes, subject matter, languages, and modus operandi—depict Chinese characters who only intermingle with other Chinese characters located within the Chinese community and who only speak Mandarin or an amalgamation of Chinese dialects such as Cantonese, Hokkien, and Hakka. This collective usage of "Chinese-ness" goes beyond its cinematic representation and is by no means accidental. It is a direct challenge and criticism of the political economy of Malaysia and its mainstream cinema. Throughout these films, this challenge is carried out through scenes of eating, which are in essence bodily representations of the need to satisfy a hunger—a desire and longing for equal identification as Malaysians irrespective of race and ethnicity.

Malaysian Sinophone cinema replicates certain characteristics of accented cinema. While "accented cinema"[1] consists of diasporic, exiled, and ethnic Third World filmmakers driven by a desire for their homeland through a character's journey toward home, the Malaysian-born Chinese Malaysian filmmakers are neither exiled nor living abroad. These filmmakers are of the opinion that they are being deprived of a sense of nationhood when they become marginalized by race- and ethnicity-based policies. Their films attempt to highlight the inequality caused by certain state policies that provide courses of preferential treatment solely for one ethnic community. Such denial of opportunities for other ethnic communities has led to a common perception that certain Malaysian citizens are of "second-class" status.[2] In the past, Chinese Malaysian filmmakers were denied state funding and their films were deemed "not Malaysian" because they did not employ Malay-centric themes and subject matter. This negation has deterritorialized Chinese Malaysian filmmakers into an "exiled" state. In response to such "exile," Chinese Malaysian filmmakers highlight their ineffectual search for Malaysia through their portrayals of characters who eat to satisfy their hunger for a sanctuary. They eat to satisfy their desire for a nation. The characters in Malaysian Sinophone cinema eat to satisfy their desire to be recognized as Malaysian.

The Malaysian Sinophone Cinema

Since its inception, Malaysian cinema has been popularly known as "Malay cinema." This is because, despite being pioneered by Malays, Chinese, and Indians, Malaysian national cinema (which is also Malaysia's commercial mainstream cinema) has primarily catered to the Malay community, focusing primarily on the lives and problems of Malays.[3] The bulk of Malaysian film narratives concentrate on Malay culture, use Malay casts speaking Malay, and are conventionally built on stereotypical melodramatic storylines revolving around elements of "*suka*" (love), "*duka*" (sadness), and "*jenaka*" (humor). In effect, Malaysian cinema is an exclusively Malay-centric industry, and because these films fail to acknowledge the diversity of Malaysia—the presence of Chinese, Indians, and other indigenous populations that constitute almost forty percent of the Malaysian population—they lead audiences to reduce Malaysian cinema to "Malay cinema."[4] Furthermore, this reduction leaves audiences with a false perception that Malaysia is made up only of Malays or that only the problems of Malays are important enough to be highlighted in films.[5] This ethnocentricity is being resisted by a group of prominent Chinese Malaysian filmmakers through Malaysian Sinophone cinema.

Three practices characterize Malaysian Sinophone cinema: its use of Mandarin and Chinese dialects to depict stories of the Chinese Malaysian community, its position challenging state policies, and its regional alignment with other Asian Sinophone cinemas. Several directors are at the center of this cinema: Tan Chui Mui, Ho Yuhang, James Lee, Liew Seng Tat, Khoo Eng Yow, and Chris Chong. They are among a number of award-winning independent filmmakers focusing on stories about the Chinese community whose films feature Chinese through dialogue in Mandarin and other Chinese dialects and through on-screen print. Their use of dialogue and print documents the linguistic complexity of Malaysia and other countries through the use of titles, subtitles, intertitles, blocks of text, and on-screen typography. As well, indicating and preserving this linguistic characterization allows these films to be more dynamic and cosmopolitan because they speak to audiences inside and outside of multilingual Malaysia. Furthermore, this practice of preserving linguistic complexity is coupled with maintaining film titles in their original language with the option of translating them into English or calling films by their English-language titles, which is an accepted convention in world cinema. (Chapman 2006) [6] For example, *If It's Not Now, Then When?* (2012) is also known as 如果还有明天 (Lit: *Ru Guo Hai You Ming Tian*), and Tan Chui Mui's *Company Of Mushrooms* (2006) is also known as 蘑菇兄弟们 (Lit: *Mo Gu Xiong Di Men*). Again, these practices challenge the Malay hegemony that has excluded other ethnic communities from Malaysian cinema. Malaysians converse in other languages besides *Bahasa* Malaysia, and such polylingual use of Mandarin and various other vernacular dialects with *Bahasa* Malaysia in films challenges the FINAS regulation that once required a film to have seventy percent *Bahasa* Malaysia dialogue for it to be recognized as "Malaysian" and eligible for a twenty-five percent tax rebate.[7] This cinema then transnationally appeals across the ethnic divide through the common use of *Bahasa* Malaysia, Mandarin, and English subtitles to compensate for the lack of understanding of a certain language.

This increased access to Malaysian Sinophone cinema has allowed it to travel extensively and win a number of accolades at international film festivals. The success, though, has led to even more discussions in the press, within academia, and on the Internet about how these films, despite winning recognition under the Malaysian flag, continue to face difficulties in securing state funding and domestic recognition by FINAS as Malaysian films. However, on January 29, 2011, then Minister of Information, Communication, and Culture Rais Yatim announced that all Malaysian-made films using Mandarin, Cantonese, or Tamil with *Bahasa* Malaysia subtitles would be recognized as "local movies" and no longer "foreign movies" as long as more than half the shooting takes place in Malaysia and more than half the production company is owned by a Malaysian. The first film to benefit from this shift in official policy was Ah Niu's Sinophone film *Ice Kacang Puppy Love*. The film was shot in Penang and stars the internationally acclaimed Chinese Malaysian singers Gary Chaw, Fish Leong, Eric Moo, and Angelica Lee, yet initially it was denied a tax rebate. After the change in policy, the film was given MYR 740,000 (US $246,667) in tax rebates. As well, other Sinophone films were also reclassified after the fact as "Malaysian" by former FINAS director general Mahyidin Mustakim who stated, "The local film industry has somehow recorded several achievements in garnering awards at international film festivals. For example, *Flower in the Pocket*,

Love Conquers All, and *As I Lay Dying* (2006) have also brought honor to the country at the international festivals like *Pusan International Film Festival (2007)* and *4th Independent Film Festival of Portugal (2007)*."[8] *In recognition of the value of these Sinophone films, the governing body began recognizing them as Malaysian films, rather than foreign films, at international film festivals.*

The success of these Chinese Malaysian filmmakers has prompted the emergence of more Malaysian Sinophone films. Some attempts have not fared as well. Bjarne Wong's horror feature *Possessed* (2006), Michael Chuah's *Seed of Darkness* (2006), and C. L. Hor's martial arts feature *Kinta 1881* (2008) were transnational coproductions, involving collaboration with companies from Hong Kong, Singapore, and China, yet they failed miserably at the box office. In 2011, though, Chiu Keng Guan's *Great Day* (2011) repeated the success of his earlier comedy *Tiger Woohoo* (2010). Both films were produced by Astro Shaw and brought in returns of MYR 6 million (US $2 million) and MYR 4 million (US $1.33 million) respectively. In 2010, Chiu also worked with singer-turned-filmmaker Ah Niu, who produced his romantic comedy *Ice Kacang Puppy Love*, which similarly did well, earning returns of MYR 4 million (US $1.33 million). Then, in 2011, controversial rapper Namewee's directorial debut *Nasi Lemak 2.0* (2011) topped the others with returns of MYR 7 million (US $2.5 million). These films were shot at locations with majority Chinese populations and make use of Mandarin and Chinese dialects. They present Chinese Malaysians mostly intermingling with other Chinese Malaysians, and build their narratives around minimalist settings and minimalist props. These filmic techniques resist existing conceptions of race, culture, politics, and national identity in Malaysian national cinema. They also demonstrate how these filmmakers have been compelled to employ transnational funding and production strategies when their films are not recognized by the state as "Malaysian" despite their films being made by Malaysians, in Malaysia, using Chinese dialects that are spoken throughout Malaysia.

In resisting Malay-centricity, Chinese Malaysian filmmakers have turned toward transnational funding and digital technology to present national issues and contest the extant political economy of Malaysian cinema. This turn has led them abroad in search of capital and coproduction opportunities and has prompted them to create a regional alignment with other Sinophone filmmakers throughout Asia. Previously, the Malaysian government has ignored these filmmakers or deemed their films "un-Malaysian," refusing to fund them because the filmmakers worked around state cinematic regulations, institutional conventions, and mainstream Malaysian cinematic traditions and practices. For example, Ho Yuhang was unsuccessful in securing state funding for *Sanctuary* (2004) because his film "was not sufficiently multicultural to represent Malaysia"[9] (Khoo 2008, 38) and not up to standards.[10] This internal neglect has pushed Chinese Malaysian filmmakers to seek support from abroad to create a body of films that have emerged as a low-budgeted transnational cinematic movement founded on transnational resource networks. Their preference for transnational networks, spaces, and strategies to produce, finance, exhibit, and distribute their films has allowed them more artistic freedom and freedom from bureaucracy.[11] Such networks have been made possible, in part, by China's rise as a global political and economic superpower and by the corresponding opening of Chinese and other markets to films and filmmakers from outside China. This opening has increasingly given many foreign filmmakers access to vast filmmaking resources and opportunities. While these Chinese Malaysian filmmakers remain Malaysians, in essence, this newly acquired access renews their cultural and economic alignment with China as Malaysian-born Chinese Sinophone filmmakers. The place of China and themes of Chinese-ness in their films have helped lead them toward a regional and global alignment with East Asian cinema networks from Hong Kong, Taiwan, and China. These Sino-nation-states have provided transnational links through coproduction opportunities, such as through Andy Lau's Focus Films from Hong Kong, as well as through engagement with new cinematic aesthetics, such as those of the Taiwan New Wave filmmakers and especially the styles of Malaysian-born Tsai Ming-liang. This form of regional alignment has also provided Chinese Malaysian filmmakers with alternative sites and platforms for the production, distribution, and exhibition of their films. These transnational networks have allowed Chinese Malaysian filmmakers greater control over their creative content, especially in higher-quality productions focusing on stories and issues related to the Chinese community in Malaysia. As these Chinese Malaysian filmmakers have reached out to these new resources and opportunities, they have also been touched by the filmmaking styles and strategies from around the region. This alignment with East Asian cinema networks has also led to the works of these filmmakers being heavily influenced by the works of art house auteurs Tsai Ming-liang, Chen Kaige, and Wong Kar-wai. This intermingling has led to the emergence of new film genres, the development of alternative narrative and aesthetics strategies, themes, and subject matter, and an evolution in cinema from many locations evoking similar representations of "Chinese-ness."

On October 15, 2013, the Chinese Film Association of Malaysia (CFAM) held the first Golden Wau Awards (GWA 2013), an event loosely based on Taiwan's Golden

Horse Festival. Malaysia's first Chinese-language film awards bestowed honors such as "Best Movie," "Best Cinematography," "Most Popular Local Chinese Film," and "Most Popular Singapore Film" on Malaysian and Singaporean Sinophone films. Among the films nominated for accolades were *Ice Kacang Puppy Love*, *Kepong Gangster* (2012), *Petaling Street Warriors* (2011), and Singapore's *Homecoming* (2011) and *We Not Naughty* (2012). This awards ceremony attended by ethnic Chinese celebrities and producers from Hong Kong, Singapore, Malaysia, and Taiwan represented local recognition of Malaysian Sinophone cinema and regional recognition within the burgeoning field of Asian cinema. The acknowledgment of these Sinophone articulations signifies a challenge to the political hegemony of China's cinema as well as a challenge to the hegemony of the dominant cinemas of their respective homelands: Malaysia and Singapore.

The Internet has also played a role in turning these filmmakers toward transnational capital. Being both cine-literate and Internet savvy has allowed them exposure to a greater flow of information and materials. Besides providing various filmmaking resources, the Internet has increased the dissemination and encouraged the propagation of notions of secularism, democracy, meritocracy, and liberal speech among these filmmakers, and has propelled them toward better understandings of notions of egalitarianism, civil rights, and social equality. This generation of highly intellectual, middle-class, well-educated, and articulate Malaysians is also bolder in expressing their critiques of injustice. They are largely an urban-based generation born after the May 13, 1969 race riots and have a deep sense of political and social consciousness, having been exposed to the socio-cultural and political conditions arising from Malay-centric nation-building policies.[12] The majority of these filmmakers were educated under the National Economic Policy (NEP) that favors the rise of the middle-class Malays or *Bumiputeras*.[13] While most of them obtained their secondary education locally, many have been denied the opportunity of tertiary education in Malaysia because of quota restrictions. As a result, they have resorted to obtaining their tertiary education abroad.[14] Living under such forms of social inequality in Malaysia has driven these directors to make state policies of affirmative action favoring certain ethnic communities the subject matter of their films.

The NEP is a policy that was initially aimed at creating equal wealth distribution in order to reduce poverty and inter-ethnic income disparity (Lee 2012, 8).[15] This was to be achieved through the restructuring of the economic disposition of national wealth by awarding economically disadvantaged ethnic communities benefits such as skills training, financial aid, and land for development. The target ethnic community was the Malay/*Bumiputera*, and the primary goal was the creation of Malay/*Bumiputera* middle-class capitalists capable of economically competing beyond national borders. This system that fundamentally denied awarding rights and benefits to non-Malay Chinese and Indians effectively encouraged a system of ethnicity-based government that racialized class formation and naturalized racial differences in the country. In effect, the NEP enhanced the political and economic capital of disadvantaged Malays at the expense of other groups within Malaysia.

Malaysian Sinophone cinema's challenge to the NEP represents one facet of a continuing desire for a Malaysian national identity that recognizes everyone regardless of race or ethnicity. Cinematographically in these films, this desire is displayed through long-takes, long shots, and narratives that develop in a deliberately unhurried pace to highlight feelings of sadness and loneliness. Such tropes emphasize the characters' (and the directors') senses of homelessness or "nationlessness," of being driven by a desire to reclaim a homeland, of deterritorialized individuals in search of their identity. Such cinematographic techniques and characterizations, in turn, reflect the experiences of the Chinese filmmakers who have experienced marginalization due to the NEP. Their predicament is highlighted through themes of loneliness, heartaches, poverty, and betrayal. For example, the characters in *Love Conquers All* (2006), *Call If You Need Me* (2009), and *At the End of Daybreak* (2009) are all longing for a sense of belonging. In these films, the Chinese Malaysian protagonists learn about the "real Malaysia" through experiences of camaraderie and love as well as through experiences of betrayal, violence, and neglect. The overarching theme of these films is desire. The characters, however, are constantly being denied what they desire. These films therefore represent the predicament faced by the Chinese community who desire Malaysia as their homeland as they are being denied equal opportunities and rights due to their ethnicity. This search for belonging represents a desire to be equally recognized as Malaysian. Since hunger is naturally fulfilled through eating, these filmmakers represent their hunger for equal recognition through scenes of eating. These scenes of eating are in actual fact a form of protest against inequality and the NEP, a form of protest challenging the strict separation of Malays and non-Malays in Malaysia and in Malaysian cinema.

The Desire to Eat and the Desire to Be Malaysian

Narratives with characters eating to satisfy their desire to belong are nonexistent in mainstream Malaysian cinema. A dining scene is often used to depict courtship and romance. Such scenes are often extravagantly portrayed regardless of their location, in a home or at a posh restaurant or five-star hotel. For example, in an overtly exaggerated dining scene in *Pisau Cukur* (2009), "gold diggers" Bella and Intan share an elaborate meal aboard a cruise ship with the affluent Datuk Zakaria. Dressed in hats and brightly colored dresses, they attempt to win Zakaria's heart over a meal. Such a use of food is in line with Malaysia's culinary reputation and popular expressions of Malaysians' gastronomic enthusiasm. In contrast, Chinese Malaysian cinema intends scenes of eating to signify something beyond courtship and romance. In this counter-mainstream Chinese Malaysian cinema, the sounds and images of hunger and eating express the desire for full national recognition and participation. They aim to challenge institutional Malaysian cinema, the NEP, and the social and political status quo in three ways.

First, in these films, eating signifies more than the basic act of consuming food to satisfy hunger. In Malaysian Sinophone films such as *Room to Let* (2002), eating goes beyond satisfying biological necessity. In fact, the characters are eating in an attempt to satisfy a desire for belonging and recognition as Malaysians. This represents the longing for equality as the NEP has been marginalizing and sidelining other ethnic communities through its courses of affirmative action. Eating thus represents the characters' attempts at sating a desire to attain equal status as Malaysians regardless of their ethnic backgrounds. Despite being marginalized and positioned at the periphery, Malaysian Sinophone filmmakers remain committed to gaining access to any form of state support, influence, pressure, and commercialization. This desire is portrayed through the films showing characters eating instant noodles rather than gastronomic meals. By showing what they are forced to eat—instant noodles are often recognized for their simplicity, cheapness, and unwholesomeness—the directors put their repression on screen. At the same time, by actually showing characters consuming instant noodles, the directors are also demonstrating the willingness of Chinese Malaysian filmmakers to make low-budget films while they are forced to continue to wait on state funds. While they wait, they continue to show the consumption of instant noodles as an unwholesome meal, highlighting their marginalization as non-Malay filmmakers.

While the lavish meals often consumed at expensive restaurants by the characters in mainstream Malaysian cinema demonstrate how state support and the NEP have helped elevate the economic and societal status of the Malay community, the consumption of instant noodles by the Chinese Malaysian characters demonstrates a state of impoverishment and neglect. Instant noodles are all they can afford. Throughout Sinophone films, characters eat instant noodles to express a sense of simplicity and as a sign of protest. The eating of something cheap signifies how they continue to be denied opportunities due to their ethnic backgrounds. The simplicity of consuming instant noodles also creates a familiarity for the audience as Malaysians, especially, are already familiar with watching television commercials featuring characters consuming Maggi Mee instant noodles and hearing the slogan: "Mi Maggi Cepat Dimasak Sedap Dimakan" (Maggi Mee, Fast to Cook, Good to Eat). As well, this slogan is always accompanied by the loud slurps of characters consuming their instant meals. Repeating such sounds and images in their films, Malaysian Sinophone filmmakers enhance local audiences' familiarity and sense of solidarity with the characters. Seeing and hearing cinematic characters eating the almost ubiquitous noodles encourages an embodied affect, a connection between audiences' and characters' desires and corporeality linked through the local. For example, the characters in the *Beautiful Washing Machine* and *Room to Let* who are lonely and broke can only afford to consume instant noodles. They too often slurp loudly as they consume the noodles. This constant consumption of only instant noodles and loud slurping is a form of protest toward the inequality of the NEP.

Malaysian Sinophone films also highlight the problems and issues faced by the Chinese community in an urbanized society. For example, *Room to Let* shows how its characters experience urbanization and the problems caused by socioeconomic inequality, embodied discomfort, psychological anxiety, and moral confusion. Their films mostly focus on the Chinese community longing for love and recognition while experiencing alienation, solitariness, and the superficiality of urban society. These films present unpretentious, unmelodramatic stories of the limitations of society, alienation of urban youth, marginalization of the working class, and dissipation through mindless capitalist consumption that allow the public to connect with these films. They see themselves or, at least, an image of themselves on screen. As such, the act of consuming instant noodles present in these films also helps audiences sympathize when the alienated characters display their sense of loss, loneliness, and sadness in an alienating metropolis. For example, the characters in James Lee's and Ho Yuhang's films often consume bowls of instant noodles in silent isolation. They sit alone at empty tables or on the couch. In *The Beautiful Washing Machine*, the reclusive main character Teoh often breaks his silence by speaking to his washing machine. The washing machine becomes his companion during meals. He treats it as a friend when he persuades it to function and wishes it good night.

This gustatory protest against inequality is often tied to silence in *A Tree in Tanjung Malim* (2004). As The Girl and The Beautiful Loser are seated at a roadside eatery, The Girl informs The Beautiful Loser that she is planning to continue her studies in Taiwan. In return, he asks her why she wants to go overseas. He asks her if the local colleges and universities are not good enough for her. She does not answer his accusing questions. In essence, they both understand her silence as an acknowledgment that it is becoming extremely difficult for Chinese Malaysian students to obtain an advanced education despite earning exam results that would warrant such opportunities. They also understand that this form of denial is due to the presence of a quota system reserving a certain number of university entries for Malay students regardless of their entry examination results. As such, rather than putting herself through the stress of attempting and failing to gain entry into a Malaysian government university, The Girl has already decided to pursue tertiary education at a Taiwanese university. Her choice of furthering her studies in Taiwan strengthens the Sinophone articulations of these filmmakers and raises issues concerning the Sinophone education networks chosen by the Chinese community.

Second, eating in Malaysian Sinophone films serves as a criticism of the restructuring of family units and human relationships. Because the nation is so often seen to mirror a certain family structure (Smith, 1991), [16] the family as the basis of a political community becomes an institution that binds its members (Honohan, 2008, 73).[17] Meals around the family dinner table present opportunities for members to communicate, bond, and learn from one another, and returning home for a meal, being able to sit together for a meal, or being welcomed to join in a meal represents family unity. In these films, such family unity mirrors national unity. In an overt criticism of the shifts of contemporary kinship structures, Long Chai in the *Elephant and the Sea* (2003) is chided for not returning home to eat with his family. While his mother grumbles that she might not have enough food to feed him and his friend Yun Ding, who turned up unannounced, they still find their place at the table and the mother serves them food as they join in the meal. The change in familial structures is also highlighted in *The Beautiful Washing Machine*. Throughout the film, silence is used to highlight the breakdown of human communication through conditions of solitude and boredom. The characters often appear wary of each other's presence in situations that warrant chattiness and conviviality. Silence is used to draw distinctions between the characters when Mr. Wong, Yap, and Yuen sit together at the dinner table. Mr. Wong eagerly waits for their approval concerning the quality of the new dish he has prepared for dinner. Despite staring at them and waiting for their feedback, he receives no compliments. The silence used during this scene and others like it demonstrates the communication breakdown within the family unit caused by modernization.

Another example of such criticisms is found in Yasmin Ahmad's *Sepet*. Openly discussing Sino-Malay relationships, her film prompts audiences to consider the value of families making time to eat together. Despite family members facing numerous problems in their personal lives, they still make time to gather as a group for a meal. During this meal, the characters converse in a variety of languages. Jason's mother speaks *Bahasa* Malaysia, his sister-in-law speaks Mandarin, and Jason's father speaks Cantonese. In these films, the dinner table represents the point of convergence for families, the time and place for everyone to gather together in an effort to highlight the importance of family values in an increasingly consumer-driven society. In *South of South* (2006), the importance of a Chinese family dining together is highlighted when the youngest family member, Ah Boy, filially invites his great-grandmother and parents to begin eating. Set in the 1980s, this sense of filial piety and importance of dining together is also further highlighted when Ah Boy is sent by his mother to call his dad home for dinner. Ah Boy searches everywhere for his father, and when he finally finds him invites him home for dinner. This working class family lives extremely simply in a wooden house with minimal, outdated furniture. One evening, an unknown figure appears at their doorstep and interrupts their dinner. The figure kneels before them and presents a gold leaf to trade for some food. The elder family members with little fuss rise from their seats, and the great-grandmother offers the unknown figure a big bowl of rice. No one speaks as the rice is accepted and the great-grandmother rejects the offer of the gold leaf. Later, the film reveals that the unknown figure is a Vietnamese refugee who has arrived in Malaysia. A short time after, the great-grandmother further reveals that these refugees have been badly treated by the authorities and that the villagers have been warned not to help them. This offer of rice in return for nothing is an image of how members of the Sinophone community might come together to assist one another. The great-grandmother's silent gesture of giving the rice depicts the tacit disapproval the Chinese communities harbor toward state policies denying them access and just participation in society. This scene also critiques the idea that because Malaysia is a gastronomic paradise and has an abundant supply of food, no one goes hungry in Malaysia. Here, again, corporeal hunger is tied to political hunger and critiques of state practices.

Third, the acts of characters from different ethnic backgrounds dining together in Malaysian Sinophone cinema represent the call to recognize the collective participation of all the Malaysian communities, including the Chinese Malaysian community, in the nation-building process. Because scenes of shared interethnic meals are unheard of in Malaysian mainstream cinema, such images challenge the conventional notion of cinematic Malay-ness for mass audiences. As a matter of fact, it is also fairly uncommon for individuals from the Chinese and Malay communities to dine together because their religious and culinary differences remain important markers of their ethnic identification. There have been instances when certain Malays have refused to share utensils with Chinese individuals due to Islamic food prohibitions that deem the Chinese consumption of pork as *haram* or unclean.[18] This form of ethnic and religious demarcation is broken down in Malaysian Sinophone cinema through scenes of Malays accepting Chinese individuals at their dining tables. These Malays that consciously depart from the Malays portrayed in Malaysian mainstream cinema portray the Malay individual as the Self (Malay nation) welcoming and accommodating the Chinese Other.

Such an evolution in ethnic identification through an evolution in gustatory practices occurs in *Flower in the Pocket* (2007). The film highlights how individuals might transcend religious and ethnic boundaries by eating together as Ayu's (Malay) mother looks beyond the ethnicity of Ayu's (Chinese) friends, brothers Li Ahh and Li Ohm. After the boys' mother dies, their father becomes a workaholic. They grow up without love until they find comfort and solace in their friendship with Ayu. Understanding the boys' need for love and compassion, Ayu's mother cares for them, treating them like her own children with love and guidance. The brothers often spend time playing with Ayu and dining at her home. Ayu's mother's acceptance of Li Ahh and Li Ohm goes beyond the simple act of feeding two hungry children. The boys' religious upbringings, cultural beliefs, and ethnic backgrounds do not matter to her. What matters to her is that the feeding of these two children represents something much more to them because they have been gravely deprived of food and love. Images of them dining together signify the transcendence of love, humility, and humanity over religious and ethnic boundaries. In this way, these scenes of eating together indicate the possibility of imagining a post-ethnic collective practice that might lead to further breaking down stereotypes. These scenes of ethnically separate characters eating together also demonstrate a form of mutual acceptance and recognition of the hunger for equality of the Chinese community. The dining table exemplifies a platform that allows for a pragmatic, intimate, and mundane space for the characters to seek refuge and build community.

Ho Yuhang and *Rain Dogs*

All three of these concerns with the implications of culinary practices throughout Malaysian Sinophone cinema are combined in the films of Ho Yuhang, especially in his feature *Rain Dogs* (太阳雨 Lit: *Tai Yang Yue*, 2006). Ho Yuhang is a crucial participant in the rise of Malaysian Sinophone cinema. Trained originally as an engineer, Ho taught himself how to make films. His first feature, *Min* (2003), was commissioned by local Malaysian television channel NTV7, while his other feature, *Sanctuary*, was funded by local production company Red Films and produced by James Lee's Doghouse73 Pictures. Partially owned by Ho and Lorna Tee from Hong Kong, Paperheart Production is known for its production-support of Malaysian-born Taiwanese filmmaker Tsai Ming-liang's film *I Don't Want to Sleep Alone* (2006). Shot in Malaysia, the film was initially banned locally; however, a censored version with five cuts eventually was allowed for release. Paperheart Productions is also responsible for the production of Ho's most prominent features: *Rain Dogs* and *At the End of Daybreak*. *Rain Dogs* was the end result of a coproduction between Ho's locally owned production company Paperheart Production, the Hubert Bals Fund, and of Ho's participation in the Papercut film project. He was among six filmmakers from Asia selected to produce a high definition (HD) feature film. This project, sponsored by Andy Lau's Focus Films, was aimed at producing Sinophone films that featured Chinese communities across Asia, including Malaysia, Singapore, Hong Kong, Taiwan, and China. *Rain Dogs* also garnered Ho the Best Director Award at the Nantes Three Continents Festival. Ho's fourth feature, *At the End of Daybreak*, was coproduced by Paperheart Production, October Pictures Ltd. (Hong Kong), and M&FC (Music & Film Company, Korea) and was supported by the Asian Cinema Fund, Hubert Bals Fund, Hong Kong Financing Forum, and Paris Projects. The film was shot entirely in Malaysia with assistance from the Taiwanese sound crew 3H. Post-production work for the film was carried out in Malaysia, Taiwan, and South Korea, and the soundtrack of the film was composed by Korean Cho Seong-woo. The film also employs a cast made up of Malaysian and Hong Kong actors, namely Tsui Tin Yau and Kara Wai Ying Hung. Ho's treatment of eating is central to understanding the significance of gustatory scenes in Malaysian Sinophone cinema.

Rain Dogs highlights the significant relation between cinematic culinary practices and the difficulties and predicaments faced by Chinese Malaysians. The film concentrates on the desire of the Chinese Malaysians to fully participate in Malaysian society though they remain marginalized by restrictive institutional policies. The primary characters, including Tung's aunt played by Yasmin Ahmad, are predominantly ethnic Chinese. As with the previous films, the use of Chinese characters speaking Mandarin and Cantonese challenges the dominant Malay-ness of Malaysian cinema. In the film, the main character, Tung, is on a journey of self-discovery and learns about the "real world" through experiences of camaraderie and love, betrayal, violence, and neglect. Tung journeys to the city in search of his estranged brother, Hung, who is later murdered in an illegal snooker hall. Things do not improve for Tung when his mother continues dating a parasitic vagrant, Chat Kor. One morning, Tung finds his brother's motorcycle stolen. He openly accuses Chat Kor of the theft, but Tung's mother defends Chat Kor and slaps Tung across the face. Feeling humiliated and betrayed, Tung moves to his fisherman uncle's home where he faces further difficulties.

In challenging elitism in Malaysian cinema, *Rain Dogs* depicts a different side of Malaysia and its cinema. Rather than portraying the "touristic" side of Malaysia with illuminated skyscrapers, bustling shopping malls, and posh restaurants, *Rain Dogs* reveals a "sleazier" side of Malaysia. The underbelly of Kuala Lumpur features prominently in Tung's search for his brother, and this setting recalls his desire to belong. Throughout his search, he sleeps in motels, is threatened by thugs, is offered the services of a sex worker, and learns about the death of his brother. In his attempt to evade the dark world that consumed his brother, Tung retreats to more familiar rural surroundings. In the end, though, he finds himself just as trapped by the immoral forces in his hometown.

In presenting this other side of Malaysia, Ho uses minimal settings and shoots on location. It is in these settings and locations that the scenes of eating also occur. The characters are never shown dining in expensive hotels or restaurants. Instead, they eat at home or at roadside stalls, reminding audiences of their poverty and isolation. As with the other films in this study, these characters display even further their insolvency through their consumption of instant noodles, which are present from the opening of the film when Hung's gang leader answers a telephone call in the illegal snooker parlor. The placement of the instant noodle cup in the scene signifies the character's lack of means, justifying why Hung's gang leader needs to resort to illegal dealings to make money. Later, Tung consumes the noodles when he is alone in Hung's apartment, again highlighting the character's poverty. Instant noodles are all Tung can afford to eat after being robbed by a pimp in the motel. The locality and simplicity of consuming instant noodles generally emphasizes how the Chinese are located as "second-class" citizens who are unable to obtain the privileges of the elites.

After visiting Hung in the city, Tung returns home to his mother. They sit at a dinner table having a meal and are surrounded by obsolete and minimalist furniture. The house is made of wood panels. The minimalist arrangement differs from the customary lavish arrangements of furnished sets in mainstream Malaysian cinema. The scene is emptied of belongings and also nearly emptied of sounds as no one speaks. Only the clicks of chopsticks hitting porcelain rice bowls is audible. Despite the lack of dialogue, this scene criticizes urbanization and the shift in familial relations. Characters and audiences feel the emptiness left behind from Hung's trip to the city in search of a more economically viable future. He can no longer wait for assistance from the state because he knows it will never come. Also, the sound of the chopsticks on the porcelain rice bowls is another audio protest, which functions similarly to the sound of characters slurping the instant noodles they consume. The sound of clanging while eating no more than two side dishes demonstrates their attempt to satisfy their desire for equality while allowing the audience to empathize and perhaps identify with the cinematic characters.

After Tung's quarrel with his mother, he moves to his uncle's house and befriends his cousin's tutor, Hui. As Hui teaches his cousin Yuk at the dining table, Hui reprimands Yuk for earning bad grades. Tung stands at a distance, consuming a bowl of dessert his aunt has prepared. In this scene, the act of eating highlights two ideas. First, Hui scolding Yuk highlights the importance the Chinese community places on institutional education and high marks. The scene stresses the belief in a correlation between earning good grades and obtaining a well-paying job in the future. Hui reprimands Tung's cousin because she understands how Chinese Malaysians need to obtain excellent grades on their examinations to compete for the restricted number of seats available at government universities. Second, this scene also highlights the predicament faced by Chinese schools in Malaysia. Chinese schools are often denied funding and recognition due to their demarcation as Other than the national school system. Here, in this scene, *Rain Dogs* combines acts of eating, acts of studying, and acts of familial chiding to remind audiences of the ethnic hunger that lies behind such portrayals. The Chinese hunger for full participation in their society; they desire equality, recognition, and funding for the Chinese schools.

While the film's narrative does not include scenes of Malay and Chinese characters dining together, it does show actors of different ethnicities playing their parts alongside one another, especially in the scene where Tung's aunt (played by the Malay actor, Yasmin Ahmad) sits with her family at the dining table. As a soft-spoken and loving mother, she embraces Tung with maternal love. The actor demonstrates her familiarity with key elements of Chinese culture as she speaks faultless Cantonese and flawlessly eats from a bowl using a pair of chopsticks. Interestingly, Ahmad also mischievously responds to critics of her earlier films. In *Rain Dogs*, she wears only a traditional "sarong" or "berkemban" and is shown sharing intimate discussions with her husband on their bed. Her wearing of this traditional Malay costume is not only a tenacious reply to criticisms of the characters in her films dressing and behaving inappropriately in intimate situations, it is also a display of an established Malay director and actor embracing the tropes of Chinese cinema and the desires of Chinese filmmakers to change the faces of mainstream Malaysian cinema.

Conclusion

In post-2000 Malaysian cinema, the re-entry of Chinese Malaysian filmmakers through Malaysian Sinophone cinema has challenged the existing political economy of Malaysia and its cinema. These filmmakers have confronted the backdrop of a Malay-centric cinema and its political and cultural suppression. Using transnational methods, without relying on state support (or being refused any form of state support), these filmmakers have moved abroad in search of capital, distribution networks, and exhibition sites to narrate stories of the Chinese Malaysian community. Through the subject matter of their films, filmic narratives, and representations of Chinese Malaysians characters, these filmmakers have forced their voices and thoughts to be heard. They have attempted to imagine their strategies to reclaim full participation in multi-racial, multicultural, and multi-ethnic Malaysian society through the scenes of eating. These scenes of eating are metaphors for the need to satisfy a desire to belong, to be recognized as fully Malaysian, and to reclaim their positions as Malaysians and Malaysian filmmakers. This sense of hunger in their films represents how the filmmakers display their desire for equality in the face of the New Economic Policy that marginalizes them. They eat to satisfy the desire for attaining equal status as Malaysians regardless of ethnic backgrounds through the formulation of an all-encompassing Malaysian national identity.

Notes

1. Naficy (2001).
2. Gatsiounis (2005), Montlake (2008). Under Article 153, this course of affirmative action has led to the widespread belief that views this preferential treatment as a special right and was not open to discussion (Hock Guan Lee, "Affirmative Action in Malaysia," in *Southeast Asian Affairs*, ed. Daljit Singh and Chin Kin Wah (Singapore: Institute of Southeast Asian Studies, 2005), 211–12.
3. Non-Malays, however, were customarily cast in intermittent roles as extras or were significant in "concealed" roles of directing, producing, and technical support. Early films such as Seruan Merdeka (1946), Selamat Tinggal Kekasihku (1955), and Gerimis (1968) have to some extent portrayed non-Malays as lead characters.
4. The official ethnic demographic makeup of Malaysia is divided according to classifications of Malays and Bumiputeras (65.1 percent), Chinese (26.0 percent), Indians (7.7 percent), and Others (1.2 percent). Department of Statistics (2001).
5. In 1980, the collapse of the Chinese-owned studio system altered the landscape of Malaysian cinema when ownership and capital control shifted from foreign to local ownership. Despite being nationalized, the focus of Malaysian films remained Malay-centric. The affirmative action or the New Economic Policy (NEP) helped establish independent Bumiputera production companies.
6. Chapman, James. *Cinemas of the world. Film and society from 1895 to the present*. London: Reaktion, 2003.
7. The FINAS Act and National Film Policy formally recognizes a locally produced film as "foreign" and not "Malaysian" if its language content is not seventy percent in *Bahasa* Malaysia. A film also fails to qualify as "Malaysian" should it contain scenes shot mainly outside Malaysia. Failure in securing the "Malaysian" film status results in the inability to qualify for the twenty-five percent tax exemption should a film be screened in local cinemas.
8. Khoo, Gaik Cheng. 2008. "Urban geography as Pretext: Sociocultural Landscapes of Kuala Lumpur in Independent Malaysian Films." *Singapore Journal of Tropical Geography* 29:34-54.

9   Exhibition overseas would almost guarantee their films would not be censored by the Lembaga Penapisan Filem (Malaysian Film Censorship Board/LPF). This also allows their films to bypass the necessary procurement of a permit from the National Film Development Corporation (FINAS) before production is allowed to begin and a license from LPF for films to be screened at local mainstream exhibition circuits.

10 The rationale behind the riots was growing fear and suspicion of the probable ascendancy of one ethnic community over the other. The Chinese were defending their language and cultural identity from Malay domination while the Malays felt the ascending support of the community toward the largely Chinese opposition parties as a threat toward their political superiority. David Brown, *The State and Ethnic Politics in Southeast Asia* (London: Routledge, 1996), 230–31.

11 Introduced in the 1970s, the term *Bumiputera* can be literally translated as "sons of the soil." Originating from a Sanskrit word, the categorization of the term *Bumiputeras* include the Malay majority and the indigenous communities of Sabah and Sarawak. Timothy P. Daniels, *Building Cultural Nationalism in Malaysia* (New York: Routledge, 2005), 40–41.

12 Teck Tan completed his law and arts degree at the University of Melbourne and then pursed filmmaking at the Australian Film TV and Radio School in Sydney, while Ho Yuhang studied engineering at Iowa State University.

13 The social and economic readjustments of the NEP were to occur from 1971 to 1990. The NEP was to restructure the economic composition of national wealth from a 4:33:63 ratio of *Bumiputera*, other Malaysians, and foreign ownership to a 30:40:30 distribution.

14 Lee, Yuen Beng. *Transnational Malaysian Cinema* (2000 - 2012). New Forms and Genres in post-2000 Malaysian Cinema. Melbourne: The University of Melbourne, 2012.

15 Smith, Anthony D. *National identity*. London; New York: Penguin. 1991.

16 Honohan, Iseult. "Metaphors of Solidarity." In *Political language and metaphor. Interpreting and changing the world*, edited by T. Carver and J. Pikalo. New York: Routledge, 2008

17 Kiong Tong Chee, *Identity Ethnic Relations in Southeast Asia: Racializing Chineseness* (Dordrecht, Netherlands: Springer, 2010), 98–99.

CHAPTER 9

Random Acts of Sensible Violence:
Horror, Hong Kong Censorship, and the Brief Ascent of "Category III"
Andrew Grossman

Abstract

Cultural studies often focuses on the marginal, less surveilled corners of underground horror and exploitation cinema as avenues of shadowy resistance and subversion, as sites where filmmakers and audiences can engage libidinal desires rendered taboo in mainstream cinema. If we believe the arguments of Herbert Marcuse, however, such subversions are themselves products of an overlording cultural repression that grants fleeting, licensed moments of apparent transgression as part of the culture industry's totalized plan of domination. These alleged subversions are thus inauthentic, mere symptoms of the industrialized repression that, in Marcuse's view, we habitually mistake for freedom. That most horror and exploitation cinema still reiterates heteronormative tropes and expectations apparently confirms Marcuse's view that what passes for "subversion" is hardly subversive. Without wholesale revolution and un-repression, it may initially seem we cannot escape from Marcuse's cultural trap. Nevertheless, there are rare moments when stealthy cinematic subversions are authentic (and not mere symptoms of the repressive bargain) because they violate not only mainstream norms but the heteronormative expectations of exploitation cinema. Focusing on the phenomenon of Hong Kong's "Category III" films, in this chapter I explore cinematic texts that sit on the borderline between symptomatic repression and true subversion *and* texts that cross the threshold, offering politico-sexual subversions unexpected by even exploitation cinema's own audiences. While these films are unlikely to exert noticeable effects on culture at large, they do provide important clues as to how authentic subversion might surreptitiously operate within texts otherwise beholden to the rules of genre.

Keywords: Sexuality, Exploitation, Horror, Marcuse, Psychoanalysis

Even for an alleged exploitation horror, *Daughter of Darkness* (1993) obsessively and monothematically fixates on the bodily violation of its young heroine, who time and again is stripped, humiliated, and penetrated at the hands of a family whose strictly patriarchal—and by extension Confucian[1]—oppressions are manifested as outright derangement. In the grisly, "rape-revenge" climax—extreme to the point of farce—the heroine reaches her breaking point, maniacally impaling the family's brother with a pair of shears and neutering the patriarch with a volley of bullets to the genitals. The nominal sequel, *Brother of Darkness* (1994)—also drawn from a contemporaneous true-crime story—inverts the genders but maintains a sociopathic view of the Confucianist body politic. Here, the abused young hero is not raped but is genitally beaten and rendered impotent at the hands of a sadistic elder brother-in-law; the predictably brutal finale witnesses the brother-in-law vengefully impaled, slashed, and gorily bashed over the head with a television set. Such neurotic, often Oedipal bodily contestations regularly marked Hong Kong's "Category III" sexploitation boom of the early and mid-1990s, in which anxious, imperiled physical bodies clearly allegorized the body politic of pre-handover Hong Kong, soon to be dominated, recolonized, and "abused" by the Mainland's paternalistic and autocratic control.

Beyond its overarching political allegory, however, Category III's ritualistic violence more generally broaches the ways cinema uses representations of the unveiled, dismembered, or violently transformed body not only to challenge the Freudian reality principle but to negotiate psychoanalytically the impossible task of making a projected, fantastic body markedly material and thus "realistic." As long as mainstream cinema remains enthralled with realist modes of representation, the cinema may well depend on depictions of hysterical sex and violence—the body *in extremis*—to claim a semblance of materiality and argue that it is not merely an empty projection. In Marxian terms, however, such aestheticizing of disembodiment (cinematic or otherwise) is precisely the problem: knowing that overriding social taboos prevent us from realizing our true bodily desires and potentials, we engage in a terrible bargain with cinema, mistaking its farcical projections for romantic ideals and exchanging material reality for a culture industry that offers only a bogus metaphysics (i.e., the Hollywood dream world). As cinematic culture becomes an inescapable part of daily life, the projections are inevitably reified in the form of celebrity bodies that are alternately beautiful and dying, pregnant and barren, dieting and ruined. This paradoxical attempt to re-embody and render mortal the projected image becomes even more complicated when we take into account spectators' own bodies, futilely longing themselves to become famous projections ripe for reification (and, in a further paradox, deification). Yet all of these fantasies are chronically limited and, in effect, censored; perceiving and sensing only the audio-visual fantasies the culture industry and conventional morality allow us, we come to censor our own desires and aspirations by rejecting our bodies in favor of the mere *opportunity* (i.e., spectatorship) to reify realistic images touted as more "authentic" than material reality. In the thesis of the Frankfurt School and other members of the New Left, two-dimensional cinema thus becomes the ideal tool to alienate audiences' desires, disembody the pleasure principle, and redirect limited material resources to the advantage of image-makers.

If psychoanalysis is correct, the alienation of bodily desire is inextricably linked to the rise of a civilization of taboos. In the Freudian myth, the primal father dominated the women of the clan and excluded his envious sons until they rebelled with patricide. Plagued by a self-interested if not wholly rational guilt, the sons thereafter imposed upon themselves sexual restraint and moderation to curb their bodily desires and distribute female pleasures (which exist in limited supply) more equitably throughout the clan. From this proposed equitability and self-censorship were born taboos and the reality principle that transforms them into law and cultural ideology. But early prohibitions against incest and adultery were insufficient; thereafter we were indemnified by infinite other taboos, as an escalating series of libidinal prohibitions checked and balanced every new human attempt at sexual and political freedom. Because this psychic indemnification had always exaggerated perceived societal dangers, however, the reality principle it produced also exaggerated the need for repression of the body as well as the mind. Logically, democracies should have no need to overstate the subversions of the id or the dangers of the pleasure principle, but they repress nearly as well as primeval dictatorships still awaiting their fathers' falls. With modernity, the pleasure principle becomes not merely repressed but explicitly neutered by industrial culture; the new freedoms proposed by postmodernity, sadly, become likewise manipulated and contained by genres constructed by a culture industry only apparently rational and in fact more amok than any infantile id. Paranoid of the daemonic libido, society doubly ensures the proposed safety of its reality principle with what Marcuse had called an overcompensatory "surplus repression"[2]—"surplus" because this politico-legal censorship of bodily and erotic expression is unreasonably layered over the inescapable, more organic censorships of our symbolically coded dreams and guilt-ridden, patricidal history.

Society's surplus repression—its cultural expectations, coercive punishments and rewards, and paranoid censorships—becomes the thin dividing line between freedom and safety, id and ego, bodily desires and constraining taboos. For the Marcuse of *Eros and Civilization*, however, we must question Freud's pessimistic premise that subjugating and tabooing the body is a "necessary" part of the bargain of socialization. Otherwise, we will succumb to lives that are repressed, mechanistic, and ultimately vicarious, as the desires we cannot live out naturally through our bodies become badly sublimated as dreams, whether nocturnal, aesthetic, or industrially produced. Yet the struggle to re-embody what has been suppressed and made fantastic is nearly impossible. For Marcuse, sexual desire is repressed not merely for the sake of ambiguous "civilization" but for the sake of capitalism's grinding engines: "…the unrepressed development…of erotogenic zones of the body…would eroticize the organism to such an extent that it would counteract the desexualization of the organism required by its social utilization as an instrument of labor."[3] Even worse, humankind has come to enjoy its own erotic repression, aestheticizing it in the form of masochism and clinging narcissistically to its own neuroses, which are always too fascinating to exchange for simplistic bodily pleasures.

If we successfully suppress our ids and stave off any unsettling returns of the repressed, we are, as Marcuse says, rewarded with a mediated "culture" that can be lived, albeit ideologically and coercively. Under the aegis of the reality principle, the individual is "geared to a rationality imposed upon [him] from outside…"[4] and, acting under mass culture's paternalistic rule, "lives his repression freely as his own life—he desires what he is supposed to desire…."[5] (Or, if you prefer, he sees the films he is supposed to see.) Thus does the reality principle make freedom and repression indistinguishable, for every perceived free content is lived within a form of strict bodily regulation and ingrained patriarchal repression, whether Abrahamic or, for our purposes, Confucianist. As we exchange unknown and true (or natural) freedoms for known and false (or societal) freedoms, we persist not under what Peter Sloterdijk has mordantly called an "enlightened false consciousness"[6]—because the *hoi polloi* may not even subscribe to Enlightenment rationalism—but merely a "functional false consciousness," much as functionally illiterate citizens can perform minimally according to the conventions of a given reality without being able to either unrepress their pleasures or decode reality's signs.

Yet if we submit to living our post-industrial alienations as highly rationalized—or even unrationalized—cinematic pleasures, society will barely let us have even those, for surplus repression, codified as cultural or political censorship, returns with a vengeance, condemning non-reproductive sexualities, delegitimizing protest, demanding consumerism, and paradoxically encouraging both bodily self-denial and materialist acquisitiveness. Between tribal warriors' exaggerated phallic sheaths and the absurd anatomical incorrectness of a child's plastic doll lie the traces of an unnecessarily bizarre history—humor, paradoxically, becomes not only the antipode of censorship but its end result. In fascistic scenarios, the censorships are unequivocal; in liberal-democratic ones, the censorships are more intellectually insulting, for legal and municipal authorities never bother to claim truths but merely reconcile arbitrarily inherited prohibitions with slowly evolving "community standards" of bodily conduct.

We thus return to the phenomenon of Hong Kong's "Category III" films, which, with their creation in 1989, exploited the liberality of a new film rating system to explore previously disallowed social, political, and/or sexual concerns.[7] In Category III films, Hong Kong audiences were no longer regaled with the impregnable and sanitized bodies of idealized celebrity culture; bodies were now seen naked, invaded, and terrorized, emblematic not of reified celebrity aspirations but of despair and nihilism, and usually (though not always) portrayed by second- or third-rate actors.[8] As a category of censorship—as all rating systems effactually are—Category III's relative permissiveness affirms the thesis of Annette Kuhn, who abandons the essentializing notion that censors interfere with a pure, unconstructed reality. Arguing along familiar lines of 1980s anti-essentialism, Kuhn posits a "productive" model of censorship that sees the censorship not as a direct intervention but as part and parcel of the construction and consciousness of early film grammar.[9] Kuhn's logic, borne of early postmodernist optimism, posits a reality principle that we actively construct rather than receive paternalistically; material reality is thus something we can organically control and transform, a sanguine notion that even Marcuse momentarily embraced in his *Essay on Liberation* (1969), which framed the counterculture and the riots of 1968 as possible turning points in the history of humankind's patriarchal domination.

Yet we cannot gloss over the real problem: who, exactly, is the "we" that is happily constructing reality? Even in Kuhn's account, the "we" is composed of filmmakers and artists, who trust that alienated audiences will provide tacit consent for filmmakers' constructed (or projected) realities. The Marcusian position on such cultural constructions, over the course of *Eros and Civilization*, *One-Dimensional Man*, and his more hopeful, post-1968 essays, ultimately balances the overriding reality of surplus repression with the rising possibility of an actual anticapitalist revolution that might topple the present dictates of the reality principle. Marcuse realizes, however, that actual revolutions, even after the ascent of post-1968 leftism, are likely to be replaced with counterfeit aesthetic revolutions easily appropriated or even manufactured by the culture industry. Capitalism's evil genius can commodify its own opposition, and civilization's discontented audiences can and will be appeased by small aesthetic subversions when true political subversions seem too overwhelming or impossible a goal.

Furthermore, within Marcuse's framework, non-creative audiences who seek catharses through allegedly subversive cultural texts are only stimulated passively and are only allowed to participate within the generic sociopathy created by the culture industry's overlording surplus repression. More simply, outré films that apparently oppose the patriarchal reality principle are themselves created by auteurist patriarchs who benefit from the usual system of capitalist domination. The question thus remains: are allegedly subversive, underground films truly transgressive, or are their necessarily evanescent rebellions part and parcel of a larger system of appeasement and ultimate domination? Is it possible for a subversive content to override a conventional, commercialistic form? For that matter, can filmmakers who sincerely want to make subversive statements do so within inherited generic boundaries, or are all "transgressive" films ensconced within established genres doomed to hypocrisy, compromise, or ultimate failure?

It is the assumption of some cultural studies critics, of course, that content can potentially overcome form, and that close textual analysis, reader-response criticism, and subjective spectatorships can trump the cruel commercialism that produces potentially transformative films. Though Marcuse, in his Marxism, remains suspicious of the vagaries of aesthetics, he, too, optimistically holds out the possibility in *Eros and Civilization* that an unrepressed, pre-bureaucratic aesthetic, which assesses physical pleasures *with* rational discrimination but *without* irrational prejudice, might deliver us from the tyranny of the reality principle. But the desire to live in a non-repressive world that makes no distinction between the pleasure principle and the reality principle, between unfettered desire and societal responsibility—in short, between content and form—may remain a utopian fantasy (a possibility Marcuse clearly realizes himself). The more limited question for our present study is whether Category III's presumably transgressive cinema of penetrable, violated, and therefore "realistic" bodies can adequately redress or subvert the romanticized, sanitized, and "aspirational" bodies of mainstream consumer culture. By entertaining even this limited prospect, we become more forgiving than Marcuse, who, at his most extreme, would argue that a true transcendence would require a legitimate overcoming of the old forms and genres and not merely the placement of subversive content within (filmic) genres already sanctioned by an oppressive reality principle. Nevertheless, even if some of Category III cinema's attempted subversions fail to transcend the reality principle to Marcuse's satisfaction, they do oppose the assumptions of mainstream, commercial cinema *to a degree*, slowly if not revolutionarily effacing the oppressive conventions of sex, gender, or power emblematic of genre filmmaking.

The Category III marker pan-generically indicates a variety of films Hong Kong colonial authorities deemed socially threatening, including glamorized gangster dramas and costumed bawdry in the manner of *Sex and Zen* (1991). Because we will address films most subversive in terms of bodily performance and representation, our focus will logically settle on Category III's notorious horror subgenre, where the body is perennially and violently at stake. Though Category III was instituted in 1989, the year of Tiananmen Square, there is, notably, no direct correlation between the two phenomena; the new category was apparently created as a reaction against the previous year's censorship of nudity in imported prestige films, such as Hollywood's *The Accused* (1988) and *The Last Temptation of Christ* (1988). Nevertheless, this marginal, often low-budget, and less surveilled category made available to genre filmmakers spaces of cultural resistance, especially as the 1997 handover prompted repressed political anxieties to rise climactically to the surface.

Themes of 1997 anxiety are familiar terrain in Hong Kong cinema, of course, and we cannot naively argue that shadowy exploitation filmmaking *totally* offered "spaces of resistance" or "modes of subversion" without also acknowledging Marcuse's larger point that such subversions are themselves products of the culture industry, which appeases audiences with small crumbs of subversive pleasure as part of its overall design of control. Nevertheless, within these small "crumbs" there are moments of true political and generic subversion and not merely generic spectacles of horror violence that fallaciously pose as subversive. In Marcusian terms, these are moments when certain films' subversive content also threatens to subvert the films' overlording genre (or form), therefore becoming genuinely (not speciously or pretentiously) revolutionary. What renders some Category III films potentially subversive is not violence alone—for the most conservative and militaristic films are often the most violent—but the ways in which Category III often renders violent actors *neurotic*, thereby undoing the erotic and political authority that aggressive bodies ritualistically signify. Furthermore, the last film we will examine—1993's *Hero Dream*—doesn't merely undo the ethos of violence within genre filmmaking but transcends the heterosexist limits of the action film by deploying transsexual bodies as unexpected objects of desire, disrupting the hermetic assumptions and operations of typical genre filmmaking.

Though many scholars have debated the issue of Hong Kong film's "1997 syndrome," it was never a reification of the Western spectator, as handover anxiety cropped up endlessly in film texts, implicitly and explicitly.[10] Yet political commentary in pre-1997, mainstream Hong Kong cinema—perhaps excluding the more verbose films of Allen Fong or Evans Chan—was generally coy and hardly contentious. Political themes in commercial films were limited mainly to offhand remarks, the nonconfrontational humanism of Ann Hui, or perilously light satire, as in *Her Fatal Ways* (*Biaojie, ni haoye!*, 1991). But if the newly installed Category III manufactured subgenres that envisaged colonial Hong Kong as an anxious young body perpetually terrorized by deviant authority, it was equally unsurprising that colonialism's impending finale demanded opportunities for long-denied, cathartic expressions of violence and horror. Though Category III did become a kind of "productive" censorship in the early and mid-1990s, demonstrating Kuhn's general thesis that censorship categories can actually manufacture rather than inhibit new modes of representation, I should stress that its attempted metaphors are often crude, repetitive, and rarely progressive in the liberal-humanist sense. The films' obsessions with bodily violation tend to be as ritualized as those of the Japanese *pinku eiga* of the 1960s and early 1970s, which, as Nagisa Oshima had argued,[11] allegorized leftist dissent with materialistic images of rape and juvenile sexual rebellion. However, numerous Category III sexploitations feature ambisexual or bisexually curious plotlines that, in the mid-1990s, were more enlightened than the openly homophobic Hong Kong films that preceded them, such as Tsui Hark's *Don't Play with Fire* (*Diyi leixing weixian*, 1980) and *The Big Heat* (*Chengshi tejing*, 1988), or Sammo Hung's lamentable *Pantyhose Heroes* (*Zhifen shuangxiong*, 1991). The exhibitionistic, taboo-breaking bodies of Category III sexploitation are also generally more politically and erotically contentious than those of Hong Kong's mainstream queer films of the period, which usually advanced nonthreatening images of bourgeois romance, in the style of *He's a Woman, She's a Man* (*Gam chi yuk sip*, 1994) or *He and She* (*Jie mei qing shen*, 1994). Category III cinema reached a dense concentration between 1992 and 1994, and in 1993—the year that most conspicuously marked the pre-1997 devolution of Hong Kong cinema's production values—approximately *a third of all films* were rated III, even top box office items such as *Flirting Scholar* (*Tang bohu dian qiuxiang*, 1993) and (rather inexplicably) *Once Upon a Time in China IV* (*Huang feihong zhi si: Wangzhe zhi feng*, 1993). By the late 1990s and especially 2000s, Category III films had faded into obscurity, partly because the post-1997 film industry as a whole deflated economically, and partly because recent films must be marketable in the tightly censored Mainland,[12] where even apolitical films dealing with folk superstition and ghosts are still deemed "regressive."

Hong Kong filmmakers are generally pragmatic in their own assessments of Category III, frustrated by its gross exploitations but also eager to take advantage of whatever uncensored opportunities the industry might momentarily offer. Classically trained actor Anthony Wong—resentful of his infamously sadistic roles and probably wishing audiences would remember him for Allen Fong's *Wuniu* (1990)—has bemusedly wondered why foreigners would even be interested in Category III.[13] Meanwhile, director Herman Yau has suggested that Category III films have vital "anti-society" tendencies and that in *Ebola Syndrome* (*Yibola bingdu*, 1996) a virulent underclass antihero symbolically enacts a "proletarian revolution."[14] In *Ebola*, Wong plays one of cinema's most abject sub-proletariat losers, a homicidal illegal immigrant whose emigration saga becomes not the contrived tragedy of Alex Law's *The Illegal Immigrant* (*Feifa yimin*, 1985) or Clara Law's *Farewell, China* (*Ai zai biexiang de jijie*, 1990) but an anti-nationalistic horror farce as unlikely as Hong Kong's colonial history itself.

After cuckolding and murdering a gangland boss in *Ebola*'s opening scene, Wong flees to post-apartheid Johannesburg, where he works illegally in a Chinese restaurant, suffering penurious wages and rampant anti-Chinese discrimination. Even more repellent than the antihero of Yau and Wong's previous *The Untold Story* (which this film openly parodies), the shabby protagonist cannot even find a whore willing to fondle him, and is reduced to the ignominy of masturbating into a raw chicken breast while eavesdropping on the nocturnal rapture of his middle-class boss. When venturing into an Ebola-afflicted tribal area to purchase discount swine, Wong rapes a semi-comatose tribal woman he fails to realize is infected with the disease. After he becomes a carrier of Ebola, he not only finds time to carve his petit bourgeois employers into "African pork buns" but embarks on an infectious spree throughout Johannesburg and later Hong Kong, bloodily liquefying an oppressive reality principle with his primal, organ-dissolving lusts. For director Yau, Wong's character is an "instinctual child" and a "primitive" whose desperate, random acts of violence demonstrate that "the lower class has many ways of revenge" against a patriarchal *cum* bourgeois regime.[15] Embodying a far more deviant version of the Thanatos Freud imagined, the hero requires an alien disease to dissolve a corrupted society and personify the "primitive childishness" that Marcuse claims has been lost to capitalist systems of production. Yet only from the point of view of society is Wong's antihero pathologized; he himself relishes his rampage, much as a clown unselfconsciously enjoys violating social norms. On the other hand, the metaphor has certain limitations, and actor Wong is more circumspect. "But their revenge is helpless.... they have no knowledge and no money," Wong responds,[16] reminding us that while Category III shock may provide subversive pleasures to the primarily young, disaffected, or alienated audiences of underground horror films,[17] such pleasures are part of the very problem, for the temporary, alienated release they offer to the disenfranchised excuses and masks the reality principle's overarching oppressions.

For the Marcuse of *Eros and Civilization*, a cinematically projected revenge such as *Ebola* has little currency unless it extends beyond the safe confines of the theater and into the didactic realm of consciousness-raising. Those who champion cult-exploitation filmmaking as oppositional or subversive surely realize that such films' cathartic or blasphemous pleasures are fleeting and vanish hours after the film is over, by which time audiences have returned begrudgingly to their proletarian or bourgeois realities. Optimists might argue that *Ebola*'s political metaphors and willingness to offend are so extreme that the film attempts to transcend even the limits of its own horror genre and the reality principle that constructs genre overall. Still, though, the film's content is inevitably and necessarily trapped within the narrative form and economics of commercial (albeit low-budget) film production. Because the film's content alone cannot entirely or successfully rebel against its domineering form, audiences are left with a kind of productive failure, an image of inspiringly amoral subversion that is itself subverted by its own commercialism and exploitativeness. Insofar as Category III films are usually generic sexploitations, audiences presumably learn to be content with this sort of titillating failure, which re-engages the bodily pleasure principle only ephemerally, and which cannot truly deliver on its ostensible promise to splinter the reality principle's overarching yet disembodied edifice.

While explicit consciousness-raising is certainly rare in Hong Kong cinema, Category III horror did determinedly focus on single themes with an almost political zealousness. In the early and mid-1990s, much of Category III horror, following the commercial success of *Dr. Lamb* (*Gaoyang yisheng*, 1992), took as its subject matter contemporary true-crime stories about perverse serial killers, who come to signify deviant ids lashing out arbitrarily against the reality principle. At the same time, these horrors often highlight the destruction of the family unit, whose annihilation signifies an admittedly crude assault on the primary site of Confucian order. Yau's *The Untold Story* (*Bat sin fan dim ji yan yuk cha siu bau*, 1993) and Billy Tang's *Run and Kill* (*Wu syu*, 1993) both climax with what remains among commercial cinema's final taboos: the graphic, onscreen slaughter of young children, here at the hands of psychopathic, unreproductive loners (scenes still less alarming, though, than Rubens' *Massacre of the Innocents*). In the former film, Anthony Wong's deviant dismembers a family and grinds their bodies into filling for his meat buns, prompting an unwitting public, in a sense, to physically consume and digest itself and, by extension, liquidate its own prospects of a tenuous post-1997 future. In the latter, Simon Yam's war-scarred Vietnamese psycho burns alive a cuckolded father's young daughter, whose charred corpse the maddened father then accidentally decapitates (a detail trimmed from the Hong Kong theatrical release but present in the Taiwanese version).[18] Other horrors obsess with patricide rather than infanticide, as do *Love to Kill* (*Nue zhi lian*, 1993), *Daughter of Darkness* (*Miemen canan zhi niesha*, 1993), *Brother of Darkness* (*Titian xingdao zhi shaxiong*, 1994),[19] and innumerable others that posit sadistic patriarchs—or, in the case of *Daughter of Darkness* and *Daughter of Darkness 2* (*Miemen canan 2 zhi jiezhong*, 1994), entire patrilineal clans—who must be cathartically slaughtered.

Onto this dichotomy of infanticide and patricide is frequently layered the theme of male impotence. In the extremely violent *A Day without a Policeman* (*Moujing shifen*, 1993), Simon Yam's divorced cop must regain his lost manhood and overcome his fear of the villains' phallic AK-47s,[20] while the perennially abused male hero of the above-mentioned *Brother of Darkness* decries his powerlessness when frontally displaying his limp penis before his girlfriend (a shot censored in the Hong Kong release but present in the Taiwanese version). In some scenarios—such as those of *Dr. Lamb*, *The Untold Story*, and *Twist* (*Zeiwong*, 1995)—themes of impotence and patricide are symbolically combined when the villainous sociopath, rather than suffering the murder demanded of a truly potent primal father, is revealed as ultimately powerless or infantile when tortured at the hands of fascistic police. In the crude, rather tedious *Twist*, the torture becomes openly sodomitic and effeminizing, as Simon Yam's incorrigible, charismatic, and eventually captured thief must endure penile electroshock and anal penetration with an industrial water hose as part of police interrogation. Admittedly, such films' metaphorical power dynamics are informed by the simplistic heteronormative rhetoric and passive-aggressive gender binaries typical of commercial Hong Kong filmmaking (especially of the early to mid-1990s) and of psychoanalysis in its most primitive, groping stages. Nevertheless, the subgenre's rudimentary psychologizing and rebounding system of sadistic infanticide and vengeful patricide give rise to a convenient allegory: the almost total destruction of bodies in these films apparently attempts to rationalize the impending political scenario of a child (Hong Kong) endangered by an all-consuming father (Mainland China) whose body must be either cathartically destroyed or brutally tortured, pacified, and infantilized. Themes of patricide and infanticide rebound and mix so freely, in fact, that the rebounding itself renders the two themes nearly indistinguishable after one has seen enough of these films. Intertextually speaking, one almost forgets if the son or the father is the guilty one; what remains is a perversely "totalized" version of the Oedipal crime, in which all members are equally doomed.

The resultant "patri-infanticide" proposes a synthesis that recalls, on a rather more sophisticated level, Freud's explanation for the paradox of the Christian Trinity: the father and son must be mixed into a single (if still magically separable) entity such that the otherwise usurping son and otherwise domineering father might reconcile transcendentally and guiltlessly. If we accept that patricide and infanticide in Category III horror become nearly indistinguishable themes, then beneath Category III's unrepressed sadism and much-discussed "anxiety" lurks the optimistic hope that former colony and fatherland may confusedly coexist. Like all utopian notions, however, such optimism is the stuff of disembodiment and abstracted metaphor; much as Marcuse himself slid between the pessimism of *One-Dimensional Man* (in which humanity is condemned to alienation) and the optimism of his *Essay on Liberation*, audiences too might shift between interpretations centering on nihilistic bodily outrage or, more optimistically, on political bodies allegorically reconciled. In either case, however, these ostensibly exploitative films, filled with gory penetrations, sexual perversions, and other emblems of material desire, ultimately attempt to re-embody a disembodied body politic.

The patriarch's penetrability becomes much more subversive in Billy Tang's *Red to Kill* (*Ruosha*, 1994), a minor cult item whose true sexual object is not a passive (and in this film, cognitively disabled) female rape victim but the handsome rapist himself. Unlike the primitive, animalistic rapists of earlier Hong Kong rape-revenge thrillers, such as *The Beasts* (*Shangou*, 1980) or *Her Vengeance* (*Xue meigui*, 1988), here the rampaging id-monster (Ben Ng) is handsome, adopts a civilized façade by daylight, and in no less than five scenes is on display nude or semi-nude.[21] In one scene, the camera stares up worshipfully at his bare, flexing buttocks, perversely inviting the film's straight male demographic to identify the rapist himself as an object of sodomy. The rapist, when clad in a skin-tight unitard, later bends his head down between his spread legs as the camera gapes squarely into his prone region. By the film's gruesome, castrating buzz-saw finale, he has inexplicably shaven his head, appearing at once phallic and infantile, predatory and purified, his very body synthesizing the problematic politics of the colonial parent-child dialectic. Yet the corporeal synthesis does not last long, and as his body is climactically split apart by the saw, the young heroines symbolically reject any notion of reconciliation, preferring to see the parent divorced from the child—and to see themselves freed from any proposed unification with an oppressive and deviant patriarch. The film's unadvertised homoeroticism, moreover, transcends the usual identifications of rape-revenge horror. By forcing upon a heteronormative horror audience a disruptive, subversively sodomitic gaze—and thus creating an unexpected, surreptitious space for transcendence within the momentary experience of spectatorship—the film arguably short-circuits the overlording cultural expectations that Marcuse assumes will not be threatened or subverted. In this rare instance, a subversive content does seem to stealthily challenge its own genre or form. If the scenario's shock value borders on absurdity, it is also worth noting that many Category III films, from the wild fantasy *The Eternal Evil of Asia* (*Nanyang shida xieshu*, 1995) to the self-conscious genre deconstruction *Daze Raper* (*Mijian fan*, 1995), are openly farcical in tone, feeding a subgenre that emphasizes poker-faced comic discomfort over outright horror.

Less self-conscious but more surprising is *The Sweet Smell of Death* (*Luoming geluofang*, 1994),[22] by all conventional standards a laughably bad film that nevertheless does what supposedly outré, genre-disrupting cult films refuse to do: consider without irony the authoritarian male body as a passive object of rape. As with the aforementioned *Twist* and other politically incorrect Category III thrillers, the film unapologetically engages regressive, heteronormative passive-aggressive gender binaries, but here the old binaries eventually yield a rare, unexpected exposé of genre filmmaking's usual narratives and practices. The film's hero is a chauvinistic, misogynistic cop who mocks rape complainants but soon becomes the target of a bisexual, chloroform-using rapist. Detaining the suspected rapist for questioning, the cop demands a sperm sample and then, when the suspect refuses, proceeds to masturbate the rapist at gunpoint to procure a specimen, whereupon the rapist licks the officer's gun in lascivious defiance. After much ludicrous cat-and-mouse, the film finally arrives at its *raison d'etre* when the flabbily bodied, rather dim cop is drugged by the rapist, stripped, and sodomized. Humiliated, he hesitantly raises his gun to his desecrated temple, but there will be no simply suicidal cleansing. Nor, apparently, can there be the feminist consciousness-raising that attends the victim of the female rape-revenge formula, for after the cop shoots his rapist dead, the credits roll abruptly, even defensively, forestalling any enlightened denouement. Obviously, there is nothing progressive about this homophobic scenario as such. The rapist's death, like the hero's violated anus, remains off screen and unglamorized, as if the film cannot confront the generic ramifications of a male rape-revenge plot whose hero represents supposedly inviolable legal authority (rather than being a civilian, like Ned Beatty in *Deliverance* [1972]). Nevertheless, the continually stretching boundaries (and low budgets) of Category III do allow for the simple *existence* of such an unsellable plot, whereas mainstream Hollywood or Hong Kong cinemas would balk at the very possibility of seeing their action heroes stripped and illegally penetrated. More important, the stealthily nonheteronormative scenarios of *Red to Kill* and *The Sweet Smell of Death* do what Marcuse says the reality principle never does: ask a repressed audience (presumably, young heterosexual males) to consider an object of desire radically different from the objects it is ideologically expected to desire. In Marcuse's formulation, we live out our lives as systematic repressions because we desire generic and predictably categorizable objects; in the light of *Red to Kill* and *Sweet Smell*, we might then ask, "If we are given to desire unexpected objects within an otherwise conventional framework (i.e., cinema), does society's global repression begin to recede?"

If besieged masculinity is the dominant theme of Category III's "Confucian" horrors, other Category III films reveal a converse scheme of female empowerment by repositioning the avenging *femme castratrice* within recognized action film or martial arts tropes. Most frequently cited in this respect is queer director Clarence Ford's *Naked Killer* (*Chiluo gaoyang*, 1992), which posits a universe of man-hating lipstick lesbians who evince not penis envy but a desire to negate (rather than claim) the dangling monstrosity of masculine dominion. Often overlooked in *Naked Killer* is that the lesbian heroine's ultimate demise is prompted by a *heterosexual* transgression with a cop (again, Simon Yam) that violates her lesbian code of honor. The nether regions of Category III elsewhere rewrite Hong Kong cinema's rules of the woman warrior to allow her increasingly sexualized (if exploitative) participation. *Offense Storm* (*Nutongdang xingfengbao*, 1993) features a gang of pretty, bat-wielding castrators, while in *Rock on Fire* (*Jimi dangan zhi zhiming youhuo*, 1994) female killers tear out men's jugulars, gouge their predacious eyes, and fight with them one-on-one in the boxing ring. Marginally more enlightened, the action-rape-revenge hybrid *Passionate Killing in the Dream* (*Yunyu diliugan*, 1992) includes a kung-fu lesbian who spends considerable screen time scolding rogues for their homophobia.

While these exploitations may seem caricatured, even ridiculous, they should be understood within Chinese film traditions that typically allowed women to become martially formidable only to the degree that they adopted a masculinized persona and forsook their feminine sexuality, much like the Mulan character of Chinese folklore. The fighting woman's Mulan position can be seen in classics like Chang Cheh's *Golden Swallow* (*Jin yanzi*, 1968), whose heroine ultimately sacrifices her martial power for heterosexual feasibility; in new wave policiers like *Righting Wrongs* (*Zhifa xianfeng*, 1986) and *The Blonde Fury* (*Shijie dashai*, 1989), wherein butch, short-haired Cynthia Rothrock is an improbable heterosexual icon; in action vehicles like *Under Police Protection* (*Jinpai shijie*, 1990) and *Dreaming the Reality* (*Mengxing xue weiting*, 1991), in which desexualized, gun-wielding women dress and act like gun-wielding Chow Yun-fat; and in once-popular martial fantasies such as *Dragon Inn* (*Xin Longmen kezhan*, 1992) and *The Heroic Trio* (*Dongfang sanxia*, 1992), in which Maggie Cheung's loveliness and delicacy militate against her plausibility as a truly skilled warrior not requiring stunt doubles or special effects. Though lowbrow sexploitations, Category III's action-rape hybrids restore to women the sexualities they sacrificed for martial power, and remind us, too, of the etymological and transformational links among the words gender, genre, and generate (from *genus*). As the recasting and regeneration of genres allow for the regeneration of gender, the binary position of the Mulan woman here dissolves into possibilities for less circumscribed female expression within the "constructive" censorship of Category III. Just as Category III horrors transformed the theretofore impregnable male body into a site of penetration and frustrated aggression, so does it remove the female body from its own biological determinism, allowing female characters the fluid sexualities *and* violences generally forbidden by reality-enforcing genres.

Until this point, the Category III films we have examined posit some stealthy, unanticipated subversions within otherwise heterosexist or heteronormative frameworks; indeed, it is the films' ostensibly conservative assumptions that render their internal subversions surprising in the first place. However, there are some Category III thrillers that dispense with gender binarism altogether, contesting the reality principle not with spectacles of revisionist violence but with images of "indeterminate" sexuality that build upon the emergent queer identities in Hong Kong cinema of the mid-1990s. We can begin with director Joe Hau's[23] deliberately opaque erotic thrillers of the mid- and late 1990s, which wilfully deny an unsuspecting and presumably straight Category III audience any clear sexual object choices. *Passion Unbounded* (*Siji sharenkuang*, 1995) features a lesbian serial rapist who fixates on butch, masculine women, a deviant antihero who seduces a male-to-female transsexual, and a cast of victims who, rather than fervently resist, submit somnambulistically to the murderous pathologies of unhinged, neurotic killers. Offering continual, blue-filtered semidarkness instead of the expected female nudity, *Passion Unbounded* provides no clear sexual identifications or allegiances for audiences expecting a routine sex thriller. Rather, the film's nebulous narrative shifts its focus among characters who queerly experiment rather than identify as either straight or gay, disrupting the generic-erotic expectations of its audience, much as *Red to Kill* inserted a homoerotic gaze into an assumedly heterosexist text.

Hau's next "ambisexual" film, the barely released *Crazy* (*Siji sharenkuang 2: Nanfeng*, 1999),[24] comes across as a perverse, unsentimental variation on Julian Lee's Category III art film *The Accident* (*Xinyuan yima*, 1999), as a married yet bisexually curious taxi driver, Ah Huei, meets the man of his dreams, Michael, while adrift in a noirish sea of existential loneliness. While in bed with his distant wife, Ah Huei masturbates thinking about Michael, whom we later learn is (of course) a homicidal lunatic; here, actor Vincent Lam (as Ah Huei) commences one of numerous semi-nude scenes that apparently attempt to coax a latent homosexual gaze from a Category III audience expecting the usual sexploitation. After an improbable serial killer plot develops, Ah Huei, wanting revenge on insensitive killer Michael, asks his wife to seduce Michael's girlfriend, as if their lesbianism will transitively fortify their own distant homosexual bond. Just as Michael's lustful killings remain conspicuously off screen, so do the characters' sexualities remain in a transitory, intermediate limbo, and the plot's continual series of betrayals and double-crosses never truly allow any of the characters' potential sexual relationships to become consummated. Though sometimes opaque to the point of incoherence, *Crazy* is ultimately about the polysemous spaces between sexual identities, successfully sidestepping the essentialist humanism that has informed many mainstream Hong Kong LGBT films of the past two decades.

If many of the films discussed thus far allegorize Hong Kong's colonial dilemma in terms of Confucianist family dynamics, writer-director-actor Julie Lee's *Trilogy of Lust* (*Xuelian*, 1995)[25] dispenses with the allegory and makes a humiliated, invaded female body the site of an explicitly textual political critique. In the opening, a Mainland Chinese pimp exhorts his customers to "Worship the Chairman," while also insisting that, "In Hong Kong, capitalists are just the same as [our communist] leaders." He thereupon charges gouged prices especially for Hong Kong patrons who wish for entrance to his underground auction of female illegal immigrants, where economically desperate Mainland women sell themselves for passage to Hong Kong. Ah Chi (director Lee), billed by the middleman as "an intellectual who likes Hong Kongers," is sold to an abusive Hong Kong fishmonger (and disbarred doctor) who never climbed from poverty. After the husband sexually molests her with a limp, bloody fish symbolic of his capitalist failure, she traumatically flashes back to her childhood, when she witnessed her anti-Revolutionary parents perish at the hands of Cultural Revolutionary soldiers. (We learn, too, that her brother was murdered in Tiananmen Square, an event rarely broached in mainstream Hong Kong cinema at the time.) Ah Chi's spirits are raised when she meets Ah Un, a handsome farmer with whom she experiences gentle adultery and liberated orgasms, at which point she realizes Hong Kong's capitalist propagandas are as false as the Mainland's Maoist ones. Realizing that as an illegal immigrant she has no legal means of solving her dilemma—and not wishing to enact the murderous, sub-proletariat rampage of so many Category III losers—she follows the old courtesan aesthetic of compulsory martyrdom by jumping from a cliff, after which Ah Un faithfully follows suit. The film ends, however, with a former communist soldier and bird trapper, glimpsed for only a brief moment earlier in the film, discovering their dangling bodies while ironically singing an anthem extolling Maoist prosperity. Screaming in terror, the white birds he traps fly from his cage, a freeze frame of their bounding flight becoming the film's coda. Though its final moments pessimistically equate the heroine's suicide with uncaged political freedom, *Trilogy of Lust* nevertheless uses the generic spaces of Category III sexploitation to advance an overtly anti-communist (and, for that matter, anti-capitalist) agenda several years before respectable auteurs such as Stanley Kwan or Fruit Chan would even begin explicitly broaching political themes.

Many of the Category III films discussed in this chapter challenge certain of the reality principle's expectations, generic conventions, and symptoms, mobilizing previously forbidden images of the unveiled and imperiled body to emphasize that Hong Kong's very existentiality, too, must finally be undressed and exposed, stripped of the anachronistic and bullying rhetoric of Confucianist nationalism. Nevertheless, Marcuse, in his more utopian modes, would urge us to go even further, to subvert the reality principle to such an extent that the subversion itself could not be framed within any single genre (such as horror or the sex thriller) or appropriated by dominant cultural forces. While critics influenced by Lacanianism or culturalist zeal may champion the psychic liberations of cinematic "pleasure," for Marcuse *generic* pleasure merely masks our overriding cultural-industrial oppressions. Without shattering the *genres* of the reality principle, even our most savage, sybaritic pleasures can be nothing more than prearranged narcotics. But because the reality principle is itself a master genre subject to political and cultural censorship, the breaking of genres should ideally reframe reality, restoring us, as Marcuse suggests, to an unrepressed, unalienated society of Orphic and Narcissistic pleasures, in which we can experience the "joy and fulfillment" of a pre-civilized state and in which the self is wholly re-eroticized.[26] The limits of our generic reality should tear apart, giving way to a Nietzschean revaluation of all values, a world that ideally places *side-by-side* the formerly sacred and profane, the beautiful and the abject. But the word "ideally" is the rub—we know too well from the past decades' throes of postmodernism how easily the culture industry absorbs, repackages, and eventually negates attempted oppositions and pseudo-subversions, particularly those of the cinema, which are predicated on marketable style far more than toxic contents resistant to appropriation. Even Marcuse's own examples of "healthy" Orphism and Narcissism in *Eros and Civilization* are only literary (and elitist) possibilities and only theoretically threaten the reality principle's overlording genres and censorships. If the problem with newly subversive genres is that they are inevitably absorbed into existing hegemonic frameworks, the film that will most violently and abruptly break with the reality principle will not be one that merely advocates novelty or shock—shocking content becomes wearisome anyway—but one that by its very nature can *never* be culturally assimilated or synthesized into existing genres. If the reality principle presently defines what is rational, such a film would have to be *irrational on its own generic terms*—unlike, for example, Dada or abstract expressionism, which employ irrationality as a tool for arguably rational ends. While the notion of a film *totally* irrational on its own generic terms seems improbable, Category III has fortunately provided one such unique film, director Lau Keung-fu's Hero Dream (*Yinyao haoqing*, 1993),[27] a low-budget oddity that includes within its narrative two antithetical, irreconcilable, yet still recognizable genres *without attempting to resolve or synthesize them*, comically or otherwise. According to the film's posters and video box covers, audiences should expect a generically macho Hong Kong action film. However, the film is unaccountably interpolated with—in addition to a few prosaic scenes of heterosexual rape and one sequence featuring a nude male bodybuilder—explicit, lengthy, X-rated sex scenes between male-to-female transsexuals equipped with both penises and breasts. The nominal plot involves cop Chin Siu-ho, hero of *Mr. Vampire* (*Jiangshi xiansheng*, 1985) and countless mainstream B-films, journeying to Thailand to avenge the death of his wife at the hands of Thai gangsters. There he joins forces with the machine gun-toting "transsexual" gang and occasionally lounges on a bed while transsexuals have tender intercourse featuring full-frontal nudity. One of the unnamed transsexuals falls in love with the hero secretly and comes rushing to his rescue wielding an automatic rifle when he is overpowered by the villains. Taking a fatal bullet for his beloved, the transsexual dies in Chin's arms, a gesture that unintentionally parodies both heterosexual tragedy and the dying embrace of male "buddy" vehicles.

Though the film is undoubtedly transphobic (the *yinyao* of the title can be translated as "flirtatious freaks" or "flirtatious monsters")[28] and frames Thailand as a site of the exotic other, the film, in its textual audacity, dispenses with the coy transgender games of Tsui Hark's *The East Is Red* (*Dongfang bubai: Fengyun zaiqi*, 1993) and other cross-dressing *wuxia* scenarios of the early 1990s. Rather than engaging in a costumed discourse of temporary gender confusion and barely implicit homoeroticism, *Hero Dream* makes explicit through naked representation the bodies that should have been unmasked all along—but which, under the reality principle, could only be mystified under layers of a beautiful, body-denying charade. In *Hero Dream*, interpretation and aesthetic subterfuge are no longer required, and transgressive sexuality emerges not through a decoding of disembodied subtext but through explicit, uncensored bodily presentation. The repressed vision of "monstrous" sexuality has not only surfaced without hermeneutic effort but has been placed in direct, clashing opposition with the reality principle that generic, heterosexist action films signify. (Mis-)directed at two conflicting audiences, *Hero Dream* presents a thesis (macho action film) and antithesis (transsexual pornography) that sit dissonantly unrequited, resulting in—depending on your viewpoint—a disorienting Eisensteinian clash, a Brechtian alienation effect, or something like the old surrealist game of exquisite corpse, in which multiple authors contribute to a single text without knowing of the other authors' contributions. Unlike Eisenstein and Brecht, however, the disorientation is not limited to a brief montage event or a displaced song—the disorientation *is the entire film*.

The film's denial of reconciliation and generic synthesis is its unique if unintentional triumph. While the reality principle and culture industry are in the business of absorbing and digesting attempted subversions, Hero Dream presents a dominant, formulaic element and a marginal, subversive element that ideologically refuse to be absorbed into one another. Here the two elements sit side-by-side, the subversive, queer half unrepressed by the normative, rigid half, the anarchic id visible in the same frame with the ego, the other in bed with the self. In *One-Dimensional Man*, Marcuse observes that this kind of irrational "unification of opposites" is typically a tool of Orwellian oppression: the prevalence of oxymoronic language in commerce and politics discredits language altogether, rendering it defunct and powerless to transcend an oppressive culture industry. Marcuse's example of an advertisement for a "Luxury Fall-Out Shelter"[29] ably illustrates how commercial language can sabotage the dialectical process. Rather than "luxury" and "fall-out shelter" acting as a thesis and antithesis that can collide into a synthesis, here the terms sit side-by-side untransformed, offering no opportunity for progress because, within the present state of the reality principle, a luxurious apocalypse has actually become a foreseeable or desired outcome. However, forcible incongruity is also the primary technique of surrealism and, for that matter, most comedy. The reality principle unfortunately mitigates the comic element of "luxury fall-out shelter" because the idea is now *actually sellable* within current political ideology. But the jarringly unreconciled and unsynthesized genres of Hero Dream could never be sellable, for presumably no heteronormative audience would want its action film interrupted by transsexual soft porn. I would suggest, therefore, that *Hero Dream*'s surrealistic juxtaposition of incompatible genres—whether intentionally or not—turns culture's "unification of opposites" on its head, removing objects from their "natural" habitat and making them uncanny, like bedroom furniture outdoors or a crocodile in a cathedral. Once the juxtaposed genres are thus exposed, a synthesis can then occur in the minds of the audience, for when the abject and the normative are presented equivalently, side-by-side, the audience will be newly empowered to judge which side is real and which is the ideological illusion.

Admittedly, *Hero Dream* is a uniquely surreal example, and we still must ask whether the many other films discussed—particularly *Ebola Syndrome* and *Red to Kill*—sufficiently rebel against the assumptions and expectations of their own genres. There are doubtless many cultural analogies to the peculiar subject matter presented here, from Hollywood genre directors who sneakily tried to subvert the Hays Code to Soviet composers such as Dmitri Shostakovich, who encoded subversive themes within the ostensibly heroic, nationalistic model of the Stalinist symphony. Whether or not such formal rebellions are relative successes or relative failures remains debatable and inevitably depends on the radicalism of the rebellion and the ethics and conscientiousness of the rebel. Yet we can be forgiving when rebellions fall short, for all of us who are not self-sustaining hermits are to some degree complicit in the culture industry, regardless of our vehement postures or protests. Even Marcuse himself, in his later, more pragmatic moments, did not always demand absolute transcendence over inherited—but hardly static—cultural forms. In his late essay "Beyond One-Dimensional Man," unpublished in his lifetime but included in his collected papers, Marcuse admits that a total desublimation of culture could be an unreasonably utopian goal—and, in strictly Freudian terms, an impossible goal, since culture *is* sublimation. As Marcuse says here, "the desublimation which is demanded today is not an undoing of civilization but an undoing only of the archaic exploitative aspects of civilization...which fostered the false consciousness, the hypocritical morality, the administered forms of fun...and the self-righteous submission to the management of human relations in our society."[30] For our purposes, a disorientation of "archaic" forms of culture may temporarily suffice when their destruction is not at hand; indeed, an instantaneous overcoming of rules may not even be desirable, as history teaches us that spontaneous revolutions are usually doomed to organizational failure. Rather, the overcoming of forms and genres, like all political work, will be gradual, though forms do require continual and expeditious remaking, lest they fall into complacency and anachronism. If the films analyzed here still evince greater amounts of surplus repression than we would like, and if their attempted rebellions transform their genres too timidly and too slowly, it then becomes our ethical responsibility to go further, to refuse both the fleeting pleasures and the overlording ideologies of our present genres until the genres are remade.

Notes

1 I here use the terms "patriarchal" and "Confucianist" more or less interchangeably within the context of Hong Kong exploitation cinema. While Confucianism cannot be reduced entirely to patriarchal domination, of course, the family unit in the films discussed is marked by a hyperbolic masculine rule that clearly parodies Confucius's classic "five bonds," which emphasize gendered obedience. Category III films' assault on dominating fathers also parallels Marcuse's insistence on overturning the patriarchy at the heart of Freudian analysis.
2 Herbert Marcuse, *Eros and Civilization* (New York: Vintage Books, 1955), 199.
3 Herbert Marcuse, *Eros and Civilization* (New York: Vintage Books, 1955), 36.
4 Ibid., 14.
5 Ibid., 42.
6 Peter Sloterdijk, *Critique of Cynical Reason* (Minneapolis: University of Minnesota Press, 1987), 44.
7 The current Hong Kong film rating system includes Category I (general audiences), Category IIa (equivalent to an American PG or PG-13), Category IIb (equivalent to a hard PG-13 or R), and Category III (audiences eighteen or older).
8 Notably, Simon Yam, who appeared in so many Category III sexploitations following his role in John Woo's *Bullet in the Head* (1990), would eventually rise to the first rank of Hong Kong actors a decade later with his starring roles in so many Johnnie To productions.
9 That Kuhn (in her 1990 classic *Cinema, Censorship, and Sexuality*) cherry-picks films only from cinema's inchoate era, when cinematic rules were still developing, also makes her social constructionist approach a *fait accompli*.
10 For instance, see *Hong Kong Hong Kong* (*Nan yu nu*, 1983), *Police Story 3: Supercop* (*Jingcha gushi 3: Chaoji jingcha*, 1992), *Rock n' Roll Cop* (*Shenggang yihao tongjifan*,1994), *The Bodyguard from Beijing* (*Zhongnanhai baobiao*, 1994), Evans Chan's *To Liv(e)* (1991), Fruit Chan's *The Longest Summer* (*Qunian yanhua tebie duo*, 1998), etc.

11 Nagisa Oshima, "Sex, Cinema, and the Four-and-a-Half Mat Room," in *Cinema, Censorship, and the State: The Writings of Nagisa Oshima*, ed. Annette Michaelson, trans. Dawn Lawson (Cambridge, MA: MIT Press, 1993), 249–50.
12 Consider, for instance, Stephen Chow's innocuous, Mainland-friendly *CJ7* (2008), which has none of the irreverence of his 1990s films.
13 From the commentary track on the American DVD release of *Ebola Syndrome* from Discotek Media, 2007.
14 Yau, *Ebola Syndrome*, Discotek Media commentary.
15 Ibid.
16 Wong, ibid.
17 Certainly, bourgeois audiences also enjoy gory and outré horror in their failed attempts to oppose the reality principle; nevertheless, Category III's extreme violence directly appeals to the disaffected, particularly in the extraordinarily sadistic *Ebola Syndrome*, which faced censorship cuts even as a Category III film.
18 Despite Category III's professed permissiveness, the more extreme III-rated films still suffer arbitrary cuts for sex and violence; a consistent rationale for cuts across all III-rated films is elusive, to say the least, when Hong Kong and Taiwanese prints are compared.
19 In *Brother of Darkness*, the patricide is only metaphorical, but the impotent hero's killing of his evil elder brother is consistent with themes of patriarchal downfall.
20 In the film's climax, Yam's cop reunites with his estranged wife to kill the rampaging villain by anally impaling him with a protruding tree branch—the effeminizing moment is slightly trimmed in the still-censored Hong Kong version, but present in Taiwanese versions.
21 Other Category III films continue in eroticizing the rapist himself, such as *The Wrath of Silence* (*Chenmo de guniang*, 1994), a direct parody of *Red to Kill*, and *Raped by an Angel 2: The Uniform Fan* (*Qiangjian 2 zhifu youhuo*, 1998).
22 The Chinese title makes it clear that the villain uses chloroform (*geluofang*) to drug his victims.
23 Though Joe Hau Wing-Choi (sometimes credited as "John Hau") is marginal as a director, he was an accomplished art director and costume designer throughout the 1980s.
24 While the Chinese title implies that this is a direct sequel to *Passion Unbounded*, there is no relation beyond the films' shared director.

25 The English title is a misnomer; "Trinity of Lust" is more appropriate. The film exists in various soft-core and hard-core versions, including a more explicit version released in Germany; my summary refers to the standard Hong Kong release. The year 1995 also witnessed Julie Lee's *Trilogy of Lust 2*, which, bereft of politics, does feature a novel murder by staple gun and giant poisonous octopus.

26 Marcuse, *Eros and Civilization*, 146–47.

27 I have previously remarked on *Hero Dream* at http://www.brightlightsfilm.com/43/atonal.php.

28 In Mandarin, the common term for transsexuals is "ren yao." The "yinyao" of the film's title, which can be translated as "human monster," is a pun intended for the film's Cantonese-speaking audience, who would conflate the similarly sounding "human" ("yan") and "flirtatious" or "horny" ("yum"). Having said that, the term "yum" appeared regularly in Category III film titles of the 1990s.

29 Herbert Marcuse, *One-Dimensional Man* (Boston: Beacon Press, 1969), 89.

30 Herbert Marcuse, "Beyond One-Dimensional Man," in *Towards a Critical Theory of Society: Collected Papers of Herbert Marcuse*, vol. 2 (London: Routledge, 2001), 115–16.

CHAPTER 10

Drifting Eyeballs: Trans-Asian Feminine Porn Tastes and Experiences
Katrien Jacobs

Abstract

This chapter outlines a post-cinematic and trans-Asian framework of pornography consumption that allows women to browse and get turned on by miscellaneous and contradictory models of embodiment and cultural affect. It compares the reactions of Hong Kong Chinese, Japanese, and American women to scenes from a Japanese hard-core movie, A Young Wife Violated before Her Husband's Eyes, and a Hong Kong Category III movie, 3D Sex and Zen: Extreme Ecstasy. The essay shows that women regard pornography with a "drifting gaze." They can be aroused by fluctuating genres because they treat pornography as an "art of failure" and ignore the rules of arousal in hard-core graphic post-cinema.

Keywords: Pornography, Post-Cinema, New Media, Feminism, Queer Studies

Introduction

In this chapter I explore feminine eroticism and porn viewing as a type of "drifting gaze" and "art of failure" by looking at women's experiences with pornography and digital media across East Asian cultures and the North American Pacific Rim. It shows that pornography and erotic cinema are sensed and consumed within a post-cinematic and trans-Asian framework. The post-cinematic framework allows people to be immersed in pornographic overindulgence through digital media networks. They can easily select, share and reactivate miscellaneous movies produced within different cultures and media industries—either for personal enjoyment or to process them in the public spaces of social media. But despite ease of access to global networks, women use cultural values to make their selections and to explore feminine or feminist sensibilities. While they browse through multiple pornographies, they also process local entertainment traditions and inhibitions regarding sexuality and sexually explicit media.

For this project, women's tastes and attitudes were examined by means of workshops that consisted of small student groups and lesbian groups in Hong Kong, Japan, and the USA. Several segments of culturally diverse hard-core pornography and soft-core erotic cinema were screened and commented on informally and through in-depth discussions. While most hard-core pornography fosters a formulaic way of representing male and female arousal and orgasm, alternative pornography and soft erotic cinema diverges from the norm and questions hard core's basic aesthetic rules. As shown in one of the pioneering studies of hard-core pornography, Linda Williams's Hard Core: Power, Pleasure, and the "Frenzy of the Visible," the key feature of American hard-core pornography is one of "maximum visibility." Viewers are supposed to get physically aroused when exposed to detailed depictions of genitals and the sex act. In this sense it is a universalizing feature of hard-core aesthetics "to privilege close-ups of body parts over other shots, to over light obscured genitals; to select sexual positions that show the most of bodies and orgasm, such as a variety of 'numbers' of the externally ejaculating penis."[1] The externalization of male strength in "the cum shot" is one of the most obvious and literal phallocentric symbols or "weapons," which has been adopted in commercial hard-core pornography in various cultures and industries. In Japanese hard-core pornography there is an added element of inequality and/or abuse between genders. Females are almost always cast as helpless and passive partners waiting to serve the goal of male satisfaction, which often also ends with a cum shot on the female face. While there are obvious differences among hard-core pornographies and censorship regulations produced in different cultural settings, it is fair to say that the model of male climax and female objectification is the most dominant and assertive model of erotic corporate entertainment.

But how do women actually view and identify with these universalizing models of embodiment and how do they approach alternative selections or questions of cultural affinity? In this chapter I outline a post-cinematic framework of pornography consumption that allows people to browse and get turned on by miscellaneous and contradictory models of embodiment and sexual knowledge. I then focus on women's reactions to scenes from two different movies—a Japanese hard-core movie featuring Sola Aoi, A Young Wife Violated before Her Husband's Eyes, and the Hong Kong "3D porn movie," 3D Sex and Zen: Extreme Ecstasy. The Japanese movie was originally marketed for private home viewing, while the Hong Kong movie was released for the general public in movie theaters as an X-rated (Category III) movie. The Japanese hard core segment focuses on a submissive woman (Sola Aoi) who is pleasured into orgasm while being physically assaulted and mentally humiliated as a "slut and failure" by two male characters, one of whom ends the session by ejaculating on her face. The Hong Kong movie features a male character who leaves his mediocre marriage to get sexual satisfaction. He stumbles upon a powerful gender-bending character, the Elder of Bliss, who can (de)construct her own genitals and use a snake-like dildonic device as a kind of weapon. The pornographic-comedy genre of this movie differs radically from hard core and develops disjointed fantasy-narratives that diverge from the main story of marital breakdown and reunion. In detailing reactions to these two movies, chapter show that women enact an "art of failure" in regard to pornography by rewriting the rules of embodiment and critical engagement. As a new generation of educated and media-inundated women and feminists, they are in search of intense experiences of arousal, but they fail to abide by the rules of hard core. They sense and react to hard core with a wide range of contradictory attitudes (pleasure, analysis, laughter, disgust, cynicism) and also look for alternative portrayals of sexual embodiment that center on cultural values and sexual identity.

Designing the Post-Cinema Workshop

In order to examine these reactions within a post-cinematic framework, I solicited the testimonies of women in Hong Kong, Japan, and the USA by means of workshops consisting of screenings of different movie segments followed by discussion. I were interested in examining fluctuating female testimonies by focusing on an emerging trans-Asian consciousness, more specifically through discussions with Hong Kong Chinese and Japanese as well as Americans and Asian-Americans who grew up around the Pacific Rim area. Ten workshops were held over six months between

March 2012 and August 2012. I recruited a mixture of heterosexual, lesbian, and "undecided" women for each event. The groups were kept fairly small (about fifteen participants) to ensure a comfortable, casual atmosphere that would allow for in-depth dialogues and discussions.

The research project was initiated and coordinated in Hong Kong, which has a porn industry and erotic film culture of its own, but where people are highly influenced by overseas products and specifically those imported from Japan and the USA. The project then traveled to Japan and the USA to solicit reactions from groups of women schooled in their local sex cultures and forms of entertainment. In refining my research questions, women were asked to respond to and rate varying samples of hard-core pornography, alternative, queer, and soft-core pornography, as well as erotic cinema. In this way the workshops simulated a social media environment where netizens can easily select and comment on fragments of sexually explicit movie culture. Since these workshops took place in public spaces, women discussed experiences of arousal and "porn taste" rather than actually getting physically turned on by touching their bodies or through masturbation. I acknowledge there are many inconsistencies between private and public porn consumption, but I were interested in how small groups of women would publicly express local or transnational tastes and attitudes.

Most of the workshops were held in sex- and queer-friendly community spaces that would typically attract a mixture of straight and lesbian women. In Hong Kong, I collaborated with the feminist and queer organizations Women's Coalition of Hong Kong and G-Spot. In San Francisco women gathered in the Center for Sex and Culture (which is known for feminist and queer activism), while in Japan, the workshop was held in the feminist sex shop Love Piece Club. From the outset, some participants defined themselves as open-minded heterosexual women and lesbians, while others defined themselves as polysexual or "undefined."

The workshops were conducted with the help of simultaneous translators into Cantonese in Hong Kong and into Japanese in Japan, and participants used their native languages and English to share reactions at different intervals throughout the workshops. Before starting the video screenings, a twenty-minute introductory discussion set out the topic of women and pornography. Then a handout was distributed with basic notes about each of the video clips. I slightly changed the selections for each of the workshops but tried to keep them as constant as possible. Participants were encouraged to write or voice reactions during screenings and then verbally discuss feelings and reactions after each screening. While some audiences ignored the handouts and chatted easily with us and with one another, others were quiet and meticulously followed the handouts while writing extensive comments. Some of the sessions were loud and "resonant" throughout the screenings, while others were quiet and analytically focused.

The women who decided to participate in our workshops were briefed about the goals of the project in advance and also signed release forms stating that their anonymity would be preserved. Since the workshops invited participants on a voluntary basis and were intended as small group gatherings, it is fair to say that they did not survey a majority of women or female consumers. For instance, the project did not attract women from different socio-economic layers of society, nor those who would have negative associations with or moral objections to pornography. Rather, the project focused on recruitment within sex-positive student groups and lesbian groups, which would consist mostly of educated, middle-class women, but nonetheless with divergent attitudes toward pornography. Most of the women were sexually active as lesbians or straight women and were active or tentative consumers of pornography. A smaller number of women were sexually inexperienced and had rarely or never watched pornography. Needless to say, it would have been difficult to recruit more widely across different age groups or socio-economic classes of society, as it is not a common activity for women or men to publicly watch and debate sexually explicit media in these social and cultural settings. I contend, as porn consumers and discussion partners within a public space, women especially have to be confident and comfortable to reveal their experiences of sexual taste and arousal. They are also encouraged to enact an "art of failure" by sharing peculiar and sometimes embarrassing aspects of their private journeys of desire and embodiment.

This chapter also reflects on the logistics of using public spaces and digital technologies that can help people feel more comfortable and alert while processing embodiment and arousal. Since we were attracting a digital generation of young women and feminists who were more acquainted with X-rated video portals, we wanted to provide an environment of over-indulgence and comfort, a space to experience physical-emotional and cerebral-ideological aspects of porn sharing. The public spaces and digital technologies necessary for basic video editing and data projection are now readily available, while the employment of smaller (computer) screens and tablet technologies gives participants some freedom to be more or less immersed in the porn scenes.

Feminine Trans-Asian Pornographies

As a research project traveling from Hong Kong to the West Coast of the USA, the project archived a wide range of individual and localized reactions, while also observing the potential of a "feminine trans-Asian" porn culture. Cultures as far apart as Australia, the United Kingdom, and China have seen a dramatic increase in the number of female consumers of pornography. A 2003 Australian online survey garnered a thousand responses and reported that seventeen percent of its self-identified users were women.[2] A 2011 survey of five thousand people carried out in the United Kingdom reported about 31.6 % of porn users were female, and specifically noted that younger women (in the age group 18–25) showed more interest in pornography when compared to older women.[3] Statistics about porn usage in China in 2011, compiled by sex researcher Pan Siuming, have indicated a similar trend that women and men of the post-"'80s and '90s generations" (those born after 1980 and 1990) have testified to watching porn in almost equal numbers.[4] Even though some of these statistics are rudimentary and still represent small samples of the population, they indicate that women's patterns of embodiment and arousal can no longer be ignored or dismissed as an anomaly.

In Europe and the USA, feminist and queer styles have meanwhile solidified into alternative porn products and culture industries.[5] There is a new wave of celebrity bloggers like Jiz Lee and Violet Blue who combine their porn productions with activist and educational initiatives. Their confident shift toward feminine or queer pornography and digital networking has reached crossover audiences consisting of male and female viewers. As these products are gaining notice within East Asian regions, there is a concomitant movement of Japanese and Korean sex products that is gingerly spreading into the West. For instance, the Japanese porn companies *Silk Labo* and *Love Cosmetic* have made efforts to produce female-friendly products featuring androgynous-looking male models and body types that are attractive to young adults around the world. In China and Hong Kong there is a growing female spectatorship of Japanese homoerotic animated cartoons and movies. The widening female fan groups of Japanese *yaoi* or Boy's Love animations have developed their own types of sexual knowledge and online behaviors. While these peculiar subcultures need to be distinguished from pornography, they are contributing to a new landscape of feminine eroticism and new media across Asia and the Pacific Rim.

Within a society influenced by both Confucian and Christian morality, Hong Kong's erotica culture is the product of various cultural norms and government regulations. Its motion picture rating system forbids hard-core pornographic films, but it has nonetheless developed a strong tradition of soft-core sex films for female and male audiences. Films showing relatively sexually explicit elements can be publicly screened as Category III films for persons aged 18 and above only. In the early 1990s, Category III films enjoyed such popularity that almost half of the screened titles at that time belonged to this category. Most of these films showcase a mixture of violence, sex, and horror. Since the late 1990s, along with the overall decline of the Hong Kong film industry, there has also been a decline in soft-core porn movies that have a theatrical release. Recently, however, there seems to be a renaissance of this genre with the top-grossing movie *3D Sex and Zen: Extreme Ecstasy* (2011), which has generated much coverage in the local media due to its mixed Asian cast and the employment of advanced cinematic 3-D technologies. The enthusiastic reception of this film by Hong Kong and mainland Chinese audiences shows that both women and men are interested in further supporting a Chinese erotic film tradition. At the same time, the movie was heavily criticized by the Hong Kong public for cheating on its mission of providing sexual gratification. While audiences generally enjoyed its local Cantonese references and sense of humor, they were unimpressed by how the movie simulated naked bodies, genitals, and sex acts, and were not aroused on by this type of "pornography."

Different from this new brand of Category III films in which erotic taste is carefully calculated and marketed to feed mainstream audiences, there is a historical erotic genre in Hong Kong known as *Feng Yue* (Scholar's Romance). Thriving in the 1970s, *Feng Yue* films were known for their keen aesthetic values and for their investigations of gender/sexual relationships. Li Han-Hsiang, the deceased *Feng Yue* director, shaped this genre with his exquisite knowledge of ancient Chinese erotica and its cultural heritage, and with his bold courage to reinvent its scenes of unbridled carnal scheming and sexual intrigues.[6] Hence, besides the Anglo-American trends toward alternative pornography, these Asian examples of erotic cinema also constitute a vital aspect of feminine porn culture.

Porn in the Post-Cinema Era

Social media environments have encouraged generations of web users to browse through and select a wide variety of erotic or pornographic products, while maintaining personal profiles and friendships around acts of sharing. In the post-cinema era, it is easier for people to collectively experience and process overindulgence while maintaining a private journey of embodiment. Wendy Chun's study of cyber culture foresaw that methods of navigating and archiving pornographic excess are driven by a Foucauldian "will to knowledge," an urge to build an idiosyncratic sexual knowledge apparatus around acts of browsing and sharing products.[7] For example, porn as the downloaded, pirated, and peer-to-peer product *par excellence* is most aptly shared among male Hong Kong students, many of whom have amassed collections as young adults before entering a university classroom. These ways of sampling products and gaining sexual knowledge have also become more common for Hong Kong women, who easily glance at and gossip about products as novelty items within their social media spaces.[8]

Susanna Paasonen has recently introduced an embodied framework of "carnal resonance" that explains how this journey may affect and resonate with people in a way that can be distinguished from older types of image technologies such as porn cinema. The starting point of her analysis is to speculate how digital objects feel to contemporary users:

> It matters how objects feel since such "feeling" gives rise to different kinds of attachment and resonance. The feel, tactility, and texture of pornography are intimately tied to its technologies of production and distribution—whether the high definition and texture of 35mm film, the grainy authenticity of gonzo and amateur videos, or the apparent immateriality of digital images, videos, and texts that consist of zeros and ones and are open to virtually endless remodification.[9]

Different types of tactility determine how we process and feel images and how we experience erotic (dis)pleasures in relation to our imaginations and our bodies. Florian Voros has further developed Paasonen's notion of resonance through ethnographic interviews with French adult male porn users. According to his analysis, porn users are vigorous consumers of pornography, who "resonate" or "re-activate" porn products by downloading, archiving, and commenting on them, and also engage in bodily techniques such as nipple-touching and breast stroking. Voros argues that pornography and its potentially clichéd scenarios do not "subjugate" users. Rather, these products are skillfully selected, archived, discussed, and gradually "domesticated" amid everyday thoughts and experiences, both imagined and incarnate:

Indeed, surfing for porn on the Internet is a highly interactive process through which we browse, click, interrupt, fast forward, rate, tag or comment on the videos we watch. Apart from moments of surfing, audiences re-produce pornography through a wide range of practices such as uploading their own photos and videos to amateur tubes and cruising websites, or more simply, by reworking mentally, while masturbating, the hottest pornographic scenarios they have recently watched. Parting from the material dimension of the activity, watching porn appears as an active and productive transformation of scripts and objects into pleasure and signification. Through these bodily practices, pornography is subjectively appropriated, transformed and altered.[10]

Even though the ideas of Paasonen and Voros about pornographic resonance are also primarily based on private experiences, they may still suggest a paradigm for observing how people may resonate erotic images and signals within public spaces. As my study shows, women are not negatively harmed or unequivocally enamored by our film segments. They are experiencing peculiar journeys of embodiment and sexual knowledge while showing fluctuating emotional states and cultural or political attitudes. They resonate sexual data while straying from the primary goal of hard-core aesthetics that is to offer (mostly male) viewers a secure path to embodiment and gratification.

Drifting Eyeballs

This project sought to find out how women in a group setting react psychologically and corporeally to porn segments within a post-cinematic framework and how their choices reflect cultural values. The notion of "drifting eyeballs" postulates that women would be open to different genres and sexual preferences without settling on an ideal choice. This openness could be due to overindulgence within digital networks and the fact that the cultural parameters of porn consumption are eroding. But the idea that women would be more flexible in their porn tastes also goes back to a series of older scientific experiments, in which sexologists have clinically measured aspects of sexual arousal and gender differences in response to sexually explicit media. The most publicized scientist of female arousal is Meredith Chivers, whose study "A Sex Difference in the Specificity of Sexual Arousal" found that heterosexual women and lesbians respond positively to a wide range of straight and queer pornographic video selections. In this tradition of arousal studies, levels of male genital arousal are compared to levels of female genital arousal in terms of changes in actual physical and brainwave responses. In her study, Chivers focused on female arousal and found that women identified more easily with varying sexual preferences, while heterosexual and homosexual men were less flexible and tended to favor one specific type of sexual or pornographic genre. Female arousal thus became characterized as more open-ended than male sexuality, with "greater intra-individual variation in preferences, behaviors, attitudes, and responsiveness to cultural influences."[11] Chivers was inconclusive about whether this type of flexibility was innate or culturally bound and technology-influenced but she seemed to favor the former explanation. Her work became widely known in Canada and the United States. While some critics appreciated her results, others criticized the categorization of feminine arousal as inborn or innate.

In our research project, we further complicated Chivers's thesis by looking at the cultural influences and technological environments of young adults and how they may stimulate experiences of (dis)pleasure. Chivers was interested in recording "actual" neuro-physiological states of arousal, but we were culling self-reported statements in a public space. I also wanted to reflect on how women expressed erotic taste and affect as cultural values. In his books *The Queer Art of Failure* (2011) and *Gaga Feminism: Sex, Gender and the End of Normal* (2012), Jack Halberstam argues that cultural values of sexual difference and dissent are interconnected as "climates" or "eco-systems." A type of dissatisfaction with conservative morality ripples into various corners of public life and legislation, and it also involves a breaking down of normative mechanisms of success, cultural growth, and economic prosperity:

> Heteronormative common sense leads to the equation of success with advancement, capital accumulation, family, ethical conduct, and hope. Other subordinate, queer, or counter-hegemonic modes of common sense lead to the association of failure with nonconformity, anticapitalist practices, nonreproductive life-styles, negativity and critique.[12]

Halberstam's art of failure differs from sectarian identity politics given that the so-called normal populations and relationship models are included within a larger trend toward sexual diversity and media awareness. Moreover, attitudes of difference and dissent can be sought through viewers' identification and dissatisfaction with mainstream media.

In our case, we sought to find out how identifiable groups of women would respond to hard-core pornography and erotic cinema. Pornography as a genre is to some extent premised on the idea that stable sexual domesticity and mediocre sex lives need to be challenged and enlivened. In this sense pornographic scripting showcases "human lowlifes" rather than morally uplifting citizens who can be loyal to their significant others and families. In the following section, first I discuss women's reactions to pornographic scripting in a typically Japanese adult video, *A Young Wife Violated before Her Husband's Eyes*, in which a woman is forcefully pleasured into orgasm and recognition of the fact that she is a "slut," a "failure" who wants a lot of sex and cannot be loyal to her boyfriend. Second, I consider female responses to a couple's sexual failure and extra-marital bliss in the soft-core comedy movie *3D Sex and Zen: Ecstasy Unlimited*. In this movie a male protagonist breaks from his marriage and experiences sexual enlightenment in a pleasure den. But he is also cast as a "double-failure" when he turns out to be a "little-endowed" lover who cannot please women altogether and who needs to undergo a penis transplant.

Female Porn Talk

"War Within Ourselves"

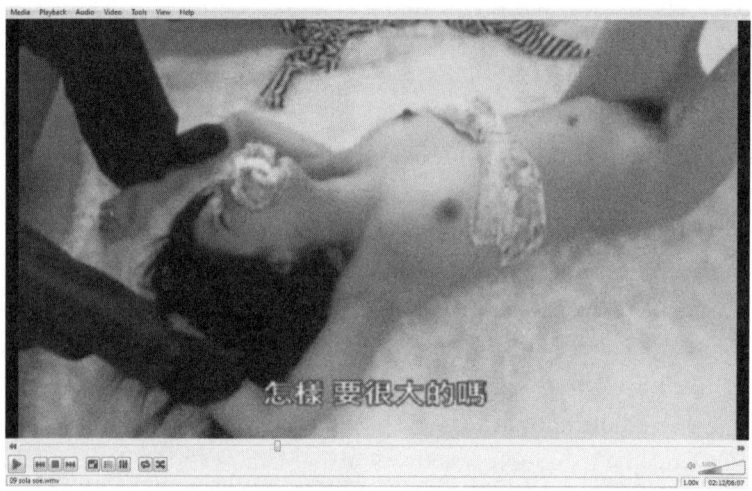

In the first clip, Sola Aoi portrays a beautiful and vulnerable young woman who is approached by two older yakuza (mafia) type men. They pull her off the couch and take off her dress and panties. The camera zooms in to show their hands roughly pleasuring her vagina and then pans to a shot of her deeply tortured face. We hear her whimpering. The men stuff her panties in her mouth and stand on her arms to prevent her from escaping. She keeps whimpering with fear and joy as she is being penetrated, first in her vagina and then in her mouth. At some point, one of the suited men pulls out a small digital camera and takes a picture of her face, in order to blackmail her boyfriend and prove her depravity, shame, and total degradation. She never speaks back to her torturers but is once again penetrated by one of them into feelings of "painful bliss." He then comes on her face and the movie is finished.

To solicit women's reactions to hard-core aesthetics, we began the screenings with a video clip featuring the Japanese pornstar Sola Aoi, who had recently become a porn celebrity throughout East Asia. Although she successfully claims the image of an empowered female entrepreneur and savvy micro-blogger, her movies for the most part do not challenge traditional hard-core aesthetics and embodiment. In many of her movies, she represents a "big-boobed youthful babe" who enjoys being treated roughly by her male companions. The movies set up a binary opposition between male and female pleasure, where male pleasure arrives at the expense of female submission. As part of a regular film plot in Japanese pornography, the woman is verbally teased or abused by her male partner and told that she is a "failure." She is accused of being a "slut" who sleeps around and cannot be loyal to her boyfriend. The male sex partner also films her while she is being verbally and physically abused, and threatens to send the video footage to her boyfriend. The scene is a typical fantasy where anxieties around moral perversion and promiscuity are projected onto the female body. Nevertheless, the young woman secretly enjoys her role of a malicious horny girl in need of punishment. She thus takes her male partner and the viewer onto a secure journey of arousal and embodiment.

Some of the Hong Kong and Japanese women hated and denounced the selected video scene because they found it very "androcentric," "disrespectful to women," or because it showed "women in a weak position." These women expressed a dislike of Japanese hard-core pornography in general. The selected clip also evoked feelings of resentment toward the dominant male actor and empathy with the suffering female. For example, as one participant explained,

> She does have a nice body, but I really want to punch the guy. When I watched the porn, I told her quietly that when the guy put the penis in her mouth, she should just bite it. And I think she agrees with me.

The American women were unfamiliar with the conventions of Japanese hard-core pornography itself. They were more distant when scrutinizing and analyzing the pornographic conventions. They were silent after the screening and then started asking extensive questions. First, they found it awkward that this clip was pixelated in order to obscure the genitals. Second, they hated the fact that the yakuza (mafia) type males in the scene remained clothed, while torturing the naked female. Furthermore, they disliked the blunt and clichéd quality of the narrative development, and especially the final "money shot" on the face of the woman.

Yet in each of the workshops, some women defended this type of scenario showing male abuse of the female, either because it aroused them or because they admired the porn star. As if taking on a "masculine" gaze, they mentioned adoring her body, her cute "anime-type" facial expressions, and her general acting talent. They specifically admired her ability to simulate a state of deep arousal, showing the dual emotions of lust and anger: "Her eyes are saying 'yes' while her mouth is begging 'no no please don't do this to me.'" Moreover, some of the Japanese women were used to depictions of female abuse and genres such as "rape pornography." One woman believed portrayals of female abuse could be arousing, even though she found this scene disappointing for its blunt repetitiveness:

> For me, when I was a young person in my teens or twenties and really interested in sex, I think this would have excited me. However, now I don't feel any eroticism from the constant pained facial expressions (gansha) or forced fellatio...but, if the situation had been more fleshed out, I might have gotten excited, even now. If there just would be a more subtle psychological characterization and rationalization for it.

A Japanese American woman summarized her reactions as a "war within herself," a mixed feeling of love and hatred toward this type of imagery. She related this directly to her mixed ethnic background, because Japanese eroticism really turns her on and "reaches her in the gut" while she also really hates it as "a strong woman raised in the USA." A Caucasian American participant reacted to this expression of internal conflict through a story of her own. She told the group that she likes one of her feminist professors but cannot agree with the professor's dislike of pornography. The clip reminded her of the fact that she has grown up feeling torn about the relationship between feminism and pornography. A third American woman echoed her statement, claiming that the scene reminded her so much of having good sex with a Japanese boyfriend who expects her to be submissive in bed. While she was not used to that kind of sex at first, she has grown to enjoy it as he has been very gentle about it.

In short the reactions among Chinese, Japanese, and American women varied. While some were angered or disliked the video, others expressed more "torn" feelings associated with enjoying Japanese erotic styles. While understanding that such fantasies can be highly reductive, some women still want to tap into them when trying to enjoy pornography and feeling sexualized. Many of the Japanese women mentioned their cultural backgrounds, either as a way of expressing dissatisfaction with hard core or more positively as a way of defending it as part of a cultural upbringing and inherited erotic tradition. American women are able to judge it from a distance, while stating that the clip is very awkward to them or pleasing to watch from an "outsider" point of view.

"It is not just a porn movie, but a porn movie made in Hong Kong"

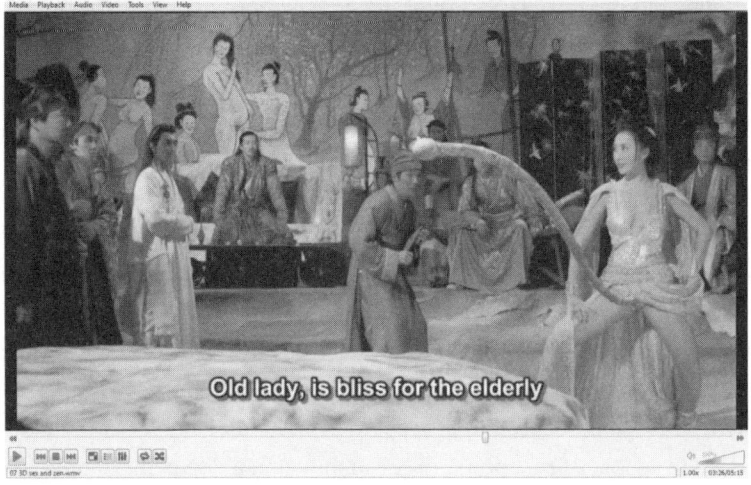

In the second clip, the protagonist Wei Yangsheng (Hayama Go) meets one of his unusual seductresses, The Elder of Bliss (Vonnie Lui), a cunning temptress and gender-fluid person who has a snake-like penis through which s/he can emulate male prowess. The Elder encircles him seductively and caresses him, while pouring a magic love potion on her bosom for him to lick. When he grabs ahold of her breast, she stretches and starts unwinding a snake-like instrument that is wrapped around her leg. When he kisses her leg, she pushes him away and stands to undergo gender transformation. Not only can she augment her breasts on the spot, she can also undo her penis and turn it into a menacing power-snake that can move independently. She uses it to impress and slap her male entourage and at one point uses it to lift and rotate a big cartwheel, while switching from female to male voice.

The movie is the third installment of a Sex and Zen series, based on a classic Chinese erotic novel from the seventeenth century, Li Yu's The Carnal Prayer Mat. Producer Stephen Shiu promoted it as entirely new genre of "3-D porn," but in actuality it is a Category III soft-core movie that caters to wider audiences and offers a hybrid of popular genres. The film begins as a comedy but ends as a very safe morality tale. It concludes by suspending its humorous tone and pronouncing the serious and highly sentimental message that true love conquers all carnal desires. The young couple who deserted each other because of their sex problems are both deprived of their genitals but are reunited through love. They both declare in a statement that sexual desire is after all not important to sustain their marriage.

The company made efforts to attract larger female audiences by introducing women-only screenings. Perhaps what is more important is the fact that the pornographic plot and the pan-Asian cast solicited spirited media debates about feminism, gender, and pleasure.

The movie follows a newly married couple who experience sexual problems. The husband, Wei Yangsheng, goes astray in a pleasure den. The pornographic plot entails a criticism of a sexually mediocre marriage bound to cause disharmony. It then shows a mixture of sexual encounters and violent adventure through which the husband tries to find sexual joy and decides to have a penis transplant to improve his performance. In the penis transplant scene, they attempt to fit him with a horse penis, but it slips off and is eaten by a dog. At that point, they can only find a dog penis to replace it. After he pleasures women with his new canine device, he provokes the jealousy of one of his rivals, who retaliates by arranging for his penis to be removed altogether.

Even though the movie focuses on the sexual pleasure and adventure of the male protagonist, Wei Yangsheng, some of the female characters are empowered with remarkable bodies or have supernatural sexual abilities. Thus we selected a scene showing the Cantonese actress Vonnie Lui as the Elder of Bliss, a supernatural character who switches between genders and employs special sex magic. They are a male in sexualized female disguise who can manipulate their female and male genitals, specifically their breasts and a snake-like dildonic device. In this capacity they are a powerful seductress for the "natural" male powers around them. Lui brings to the movie an applauded element of dazzling beauty and queer empowerment, but her character is later obliterated as they seduce a Zen master and are killed by one of their enemies when they returns from their mission and a special device that was inserted in their body explodes. As with the male protagonist, the Elder of Bliss is punished for going against natural human sexuality and natural human embodiment.

The women in our Hong Kong workshops were familiar with the movie and agreed in labeling it a male-oriented movie that does not make strong efforts to celebrate or support feminism or queer sexuality. The movie was promoted in Hong Kong as an entirely new genre ("3D-pornography") and was also branded with Hong Kong pride. It had a massive theatrical release and viewer response in Hong Kong in 2011, but was also criticized by the male and female public for cheating on its mission and "not being hard core enough." For example, many of the male viewers in Hong Kong complained about the fact that the Cantonese actress only revealed "fake" artificially augmented breasts rather than her real breasts. While Hong Kong audiences were disappointed with the film's rhetoric of pornographic bravery and technical innovation, they did appreciate its attempt to revive local Chinese storytelling and Cantonese vulgarities.

Some Hong Kong women were duly entertained and impressed with its Hong Kong quality, which is primarily conveyed through a specific type of humor, the use of Cantonese sex talk and vulgar expressions, as well as the use of ancient Chinese eroticism. These elements were appreciated as a way to build up a local sex culture that would be able to remain uncensored in Hong Kong. As some participants stated,

> You see these crazy sex scenes, like the "penis transplant scene." That kind of low-brow humor is not around anymore in Hong Kong cinema. I appreciate this.

> Also, as a Cantonese speaker, it is rare to hear those swear words in cinema, those that you hear on the street. That really adds to the enjoyment of the movie. For instance "mo lang yong" means "cock" … it also means that you are a complete loser. That word was used in the movie it was done in a very funny way. That is Cantonese humor and there is a lot of that in the movie.

> Hong Kong people are happy that we can produce this kind of movie, while it would be totally censored in mainland China. You know that many mainland Chinese people travel here to see the uncensored version. Secondly, the moralistic tone at the end of the movie is typical… to use an ancient story and it ends up as a kind of moral fable. Thirdly, even the use of a horse's penis as a transplant makes a reference to local medicine, as they make use of that body part of various animals, like a tiger's penis.

> It is a new trend here. People are more nostalgic for local movies. For many people it is not a just porn movie but a porn movie made in Hong Kong.

When we asked the Hong Kong participants to react to the selected scene of post-human fantasy, overall they said they did not find it arousing. Most of them had seen the entire movie and found it culturally significant or entertaining but not at all titillating. They also reported they were familiar with the movie's male-centric and moralistic view on sexual embodiment and were not aroused by the queer film segment. Some reported,

> The scene is just funny, but in the end it is not erotic at all. She is supposed to be a femme-fatale here, but she is not really shown in an alluring way. I think this scene even may be a bit homophobic in its portrayal of a hermaphrodite character. I think that most people would have negative associations with this kind of portrayal.
>
> For me this scene is a turn-off compared to all the other movies that we have watched. Even the seduction with the snake-like penis does not work for me.
>
> Overall the movie is just a money grabber and as a result, it does not have feminist values at all.

The Japanese and American participants were unaware of the cinematic context and general reception of the movie and were more much more aroused by the segment. The Japanese women specifically liked the fact that the post-human character can augment their breasts and manipulate a penis-like device. They stated that they would definitely go see this kind of strange fantasy movie based on the selected clip. They also stated that they were aroused by the scene because it reminded them of Japanese movie culture, specifically anime movies with hyperbolic female characters or Japanese *pink movies* from the 1960s/1970s that are currently going through a revival period in Tokyo and are being shown in designated movie theaters.

The American participants admired the scene as a "fun" and "cheesy" type of "samurai" or "kung fu" genre rather than a porn movie. They also were aroused and charmed by the scene as a good example of queer empowerment. They compared it to their own type of porn/erotica culture and the genre of spoof porn that is based on Hollywood movies or to Asian movies that are widely watched in the United States:

> I grew up with a lot of campy and silly Asian cinema and I see this as a good example of that. I am not expecting anything else from this kind of movie and I like it.

Besides locating those types of influences, Japanese and American women were unfamiliar with the issue of reclaiming Chinese eroticism through Cantonese identity. They obviously did not live through the social debates about sex and identity produced by this movie. Perhaps the reason why non-Chinese women identify sexually with the Elder of Bliss character is because they embody an ultimate fantasy figure detached from references to Cantonese movie culture. Perhaps it is easier for women to get turned on by sporadic film segments that represent a quaint type of decontextualized "orientalism." As one of the American women shouted after the screening,

> I would totally want that kind of strap-on. Next time I am going to dress up I will be thinking about that. (Laughter in the room.)

Further Analysis and Conclusions

Halberstam has postulated a gender climate where women endorse an art of failure and seek alternative models of social life, education, and entertainment. In this chapter, I have appropriated the art of failure to analyze novel feelings and practices of embodiment and arousal around feminist or feminine pornography. In Chivers' studies about women's pornographic arousal it was claimed that women easily appropriate a "drifting gaze" when confronted with multiple pornographic segments, as they easily get turned on by miscellaneous genres, sexual fetishes, orientations, or body types. My workshops confirm this claim and demonstrate that women get aroused by a wide variety of pornographic styles and segments. In this way, women's sense of embodiment is open-ended, and universalizing models of hard-core aesthetics fail to arouse all women. While some women can identify well with stories of female submission and masochism, they mostly have divided reactions and are too aware of gender politics to be simply aroused.

But women also stray from hard-core aesthetics because they want to endorse local cultural values and entertainment models when consuming pornography. For instance, Chinese and Japanese women have strong negative reactions to American queer pornography because it shows body types and sex acts they find unattractive. American women of various sexual orientations get aroused by lesbian-produced American porn but cannot endorse Japanese versions of queer or feminist pornography. In this sense, women express a desire to traverse cultural boundaries but are mostly in dialogue with their local sexuality issues and media genres.

In other words, women stray from hard-core aesthetics because movies are too formulaic and do not at all deal with cultural issues. For instance, in Hong Kong the workshops were carried out amid a climate of fierce battles over national identity and legislative autonomy from mainland China. While Hong Kong citizens crave to be a post-colonial nation in search of a unique political-legislative entity and identity, they also show a desire for a local sex culture. Their reactions to a segment from *3D Sex and Zen: Extreme Ecstasy* show their search for cultural identity also influences their sense of embodiment and arousal. They are very appreciative of the Cantonese comedy story of male incompetence as it reflects how they feel about their political leadership. At the same time they are suspicious of the film company's mission of providing female-friendly entertainment, as this movie's foremost concern is to provide male-oriented entertainment and pass the censorship board.

Japanese women differ from Hong Kong women in that they have inherited a richly flourishing male-oriented porn culture. In their case, reactions for or against a certain porn genre often stem from a grassroots feeling of female anger or jadedness, a reaction to the idea that male porn culture will never make room for female satisfaction. The emerging work of feminist companies show that the idea of an open-ended gaze is narrowly applied to the idea of feminine servitude, the idea that women have to learn how to please men, or the idea that effeminate actors can suit high-powered women. There is less room for other variations of masculine and feminine spectatorship, just as these newly born feminist companies do not endorse queer rights. This is quite different from American feminist and queer porn producers who have combined sex work with sex activism and who also comment on larger political issues. In short, the era of post-cinema will provide a new type of overindulgence alongside ongoing political tensions and divides. In this way women will use their states of arousal to love, hate, enjoy, disdain, or remain ambivalent to porn while grappling with a transformation of political or sexual freedom and associated cultural values.

* This research project was partially funded by a direct grant from the Chinese University of Hong Kong entitled "Women, Digital Media and Erotic Tastes in Hong Kong and the USA" (Grant Number: 2010374, January 2012–September 2013). I would like to thank all the institutions and individuals who helped me with this project: Love Piece Club, the Women's Coalition of Hong Kong, the Center for

Sexuality and Culture, Uma and Merdre, Kazumi Nagaike, Kaori Yoshida, Carol Queen, Library Vixen, Minori Kitahara, Yuki Kusano, Utako Shinatsuji, Kimberly De Vries. I would also like to thank my research assistants Cynthia Lam, Clara Tang, John Skutlin, and Gloria Pang.

Notes

1. Linda Williams, *Hard Core: Power, Pleasure, and the "Frenzy of the Visible"* (Berkeley: University of California Press, 1989), 46.
2. Catherine Lumby, Kath Albury, and Alan McKee, eds., *The Porn Report* (Melbourne: Melbourne University Publishing, 2008), 17.
3. Clarissa Smith, Feona Attwood, and Martin Barker, "UK Porn Research Preliminary Statistics" (2012), http://www.pornresearch.org/Firstsummaryforwebsite.pdf.
4. Meilian Lin, "Porn's Peephole," Global Times (October 30, 2011). Available at http://www.globaltimes.cn/NEWS/tabid/99/ID/681579/Porns-peephole.aspx.
5. Tim Stuttgen, ed., *Post/Porn/Politics: Queer and Feminist Perspectives on the Politics of Porn Performance and Sex Work as Cultural Production* (Berlin: B_books, 2010).
6. Yau Ching, "Porn Power: Sexual and Gender Politics in Li Han-hsiang's Fengyue Films," in *As Normal as Possible: Negotiating Sexuality and Gender in Mainland China and Hong Kong*, ed. Yau Ching (Hong Kong: Hong Kong University Press, 2010), 113–30.
7. Wendy Chun, *Control and Freedom: Power and Paranoia in the Age of Fiber Optics* (Cambridge, MA: MIT Press, 2006).
8. Katrien Jacobs, *People's Pornography: Sex and Surveillance on the Chinese Internet* (Bristol, UK: Intellect Books, 2011).
9. Susanna Paasonen, *Carnal Resonance: Affect and Online Pornography* (Cambridge, MA: MIT Press, 2011), 99.
10. Voros, Florian Forthcoming, "Domesticated porn: Gendered Embodiment in Audience Reception Practices of Pornography". In E. Biasin, G. Maina and F. Zecca, Porn after Porn: Contemporary Alternative Pornographies Milano, Mimesis, 2014.
11. Meredith Chivers, Gerulf Riefer, Elizabeth Latty, and Michael Bailey (2004), "A Sex Difference in the Specificity of Sexual Arousal," *Psychological Science* 4: 11 739.
12. Jack Halberstam, *The Queer Art of Failure* (Durham, NC: Duke University Press, 2011), 89.

Bibliography

Shen Chunhua Life Show "Zai Luofu Gong kanjian ta de lian: Cai Mingliang," TV broadcast, 43 min 49s, September 21, 2007.

Abbas, M. Ackbar. *Hong Kong: Culture and the Politics of Disappearance*. Vol. 2. Minneapolis: University of Minnesota Press, 1997.

Ahmad, Aijaz. *In Theory: Classes, Nations, Literatures*. London: Verso, 1994.

Alatas, Syed Farid. "Singapore Is Not Yet Truly Multicultural." *The Straits Times*, November 9, 2011, A14.

Althusser, Louis. *Lenin and Philosophy*. London: New Left Books, 1971.

Amer, Sahar. "Cross-Dressing and Female Same-Sex Marriage in Medieval French and Arabic Literatures." In *Islamicate Sexualities: Translations across Temporal Geographies of Desire*, edited by Kathryn Babayan and Afsaneh Najmabadi, 72–111. Cambridge, MA: Harvard University Press, 2008.

Anagnost, Ann. *National Past Times: Narrative, Representation and Power in Modern China*. Durham, NC: Duke University Press, 1997.

Ang, Ien. *On Not Speaking Chinese: Living between Asia and the West*. New York: Routledge, 2001.

———. "Together-in-Difference: Beyond Diaspora, into Hybridity." *Asian Studies Review* 27, no. 2 (2003): 141–54.

Appiah, K. Anthony. "Identity, Authenticity, Survival—Multicultural Societies and Social Reproduction." In *Multiculturalism: Examining the Politics of Recognition*, edited by Amy Gutmann, 149–63. Princeton, NJ: Princeton University Press, 1994.

Bachner, Andrea. "Cinema as Heterochronos: Temporal Folds in the Work of Tsai Ming-liang." *Modern Chinese Literature and Culture* 19, no. 1 (Spring 2007): 60–90.

Barlow, Tani E. *The Question of Women in Chinese Feminism*. Durham, NC: Duke University Press, 2004.

Barthes, Roland. *Camera Lucida: Reflections on Photography*. New York: Macmillan, 1981.

Baugh, Bruce. "Body." In *The Deleuze Dictionary*, edited by Adrian Parr, 35–37. Rev. ed. Edinburgh: Edinburgh University Press, 2010.

Bazin, André. "On the *politique des auteurs*." In *Cahiers du Cinema: The 1950s. Neo-Realism, Hollywood, New Wave*, edited by Jim Hillier, 248–59. Cambridge, MA: Harvard University Press, 1985.

Bei, Dao. *Waves: Stories*. Edited by Bonnie S. McDougall. Translated by Bonnie S. McDougall and Susette Ternent Cooke. New York: New Directions Books, 1990.

Berlant, Lauren G. *Cruel Optimism*. Durham, NC: Duke University Press, 2011.

Berry, Chris. "Jia Zhangke and the Temporality of Postsocialist Chinese Cinema: In the Now (and Then)." In *Futures of Chinese Cinema: Technologies and Temporalities in Chinese Screen Cultures*, edited by Olivia Khoo and Sean Metzger, 113–25. London: Intellect, 2009.

———. "Stellar Transit: Bruce Lee's Body or Chinese Masculinity in a Transnational Frame." In *Embodied Modernities: Corporeality, Representation, and Chinese Cultures*, edited by Fran Martin and Larissa Heinrich, 218–34. Honolulu: University of Hawai'i Press, 2006.

Berry, Chris, and Mary Farquhar. *China on Screen: Cinema and Nation*. Film and Culture (Ficu). New York, NY: Columbia UP, 2006.

Berry, Chris, and Mary Farquhar. *China on Screen: Cinema and Nation*. New York: Columbia University Press, 2006.

Berry, Chris, Lü Xinyu, and Lisa Rofel. *The New Chinese Documentary Film Movement: For the Public Record*. Hong Kong: Hong Kong University Press, 2010.

Berry, Michael. *Speaking in Images: Interview with Contemporary Chinese Filmmakers*. New York: Columbia University Press, 2005.

Bíró, Yvette. "Tender Is the Regard: *I Don't Want to Sleep Alone* and *Still Life*." *Film Quarterly* 61, no. 4 (Summer 2008): 34–40.

Bohan, Janis S. *Psychology and Sexual Orientation: Coming to Terms*. New York: Routledge, 1996.

Bordwell, David, and Kristin Thompson. *Film Art: An Introduction*. New York: McGraw-Hill, 2008.

Bourne, Christopher. "Jia Zhang-ke's 'Useless': 2007 New York Film Festival Review." *Meniscus* (2007). http://www.meniscuszine.com/articles/20071106792/documentary-review-jia-zhang-kes-useless/.

Brown, David. *The State and Ethnic Politics in Southeast Asia*. London: Routledge, 1996.

Brunette, Peter. *Wong Kar-wai*. Urbana: University of Illinois Press, 2005.

Brunette, P. *Wong Kar-Wai*. Urbana, IL: University of Illinois Press, 2005.

Buller, Rachel Epp, ed. *Reconciling Art and Mothering*. Aldershot: Ashgate, 2012.

Butler, Judith. "Imitation and Gender Insubordination." In *The Judith Butler Reader*, edited by Sara Salih, 119–37. Malden, MA: Blackwell Publishing, 2004.

"Cai Mingliang chugui qingding nanyan yancheng: women shi zhongshen banlü." China *Fenghuang wang (Shaanxi)*. September 6, 2013. http://sn.ifeng.com/yulepindao/yulequan/detail_2013_09/06/1195897_0.shtml.

"Cai Mingliang *Jiaoyou* rang 22 nian zhi'ai luniao "zhongshen banlü" gaobai." Taiwan *Pingguo ribao*. September 6, 2013. http://ent.appledaily.com.tw/enews/article/entertainment/20130906/35274422/.

"Cai Mingliang tan Li Kangsheng: Rending zhongshen banlü." *Taiwan Zhongshi dianzi bao*. September 6, 2013. http://showbiz.chinatimes.com/2009Cti/Channel/Showbiz/showbiz-news-cnt/0,5020,130511+172013090600569,00.html.

Carter, Erica, and Rodney Livingstone. *Béla Balázs: Early Film Theory: Visible Man and the Spirit of Film*. New York: Berghahn, 2010.

Chan, Kenneth, ———. "Cross-Dress for Success: Performing Ivan Heng and Chowee Leow's 'An Occasional Orchid' and Stella Kon's 'Emily of Emerald Hill' on the Singapore Stage." In "Where in the World is Transnational Feminism?" *Tulsa Studies in Women's Literature* 23, no. 1 (Spring 2004): 29–43.

Chang, Hsiao-hung. *Guaitai jiating luomanshi*. Taipei: Shibao, 2000.

———. "Taiwan Queer Valentines." In *Trajectories: Inter-Asia Cultural Studies*, edited by Chen Kuan-hsing, 257–67. London: Routledge, 1998.

Chapman, James. *Cinemas of the world. Film and society from 1895 to the present*. London: Reaktion, 2003.

Chapple, Linda. "Memory, Nostalgia and the Feminine: *In the Mood for Love* and Those *Qipaos*." In *Millennial Cinema: Memory in Global Film*, edited by Amresh Sinha and Terence McSweeney, 209–21. New York: Columbia University Press, 2011.

Chee, Kiong Tong. *Identity Ethnic Relations in Southeast Asia: Racializing Chineseness*. Dordrecht, Netherlands: Springer, 2010.

Chen, Baoxu. "Yuwang, yapo, bengjie de shengming-fang Cai Mingliang." In *Heliu*, edited by Jiao Xiongping, 52–76. Taipei: Huangguan, 1997.

Chen, Baoxu, and Tsai Ming-liang. "*Aiqing wansui* wan wan sui!" In *Aiqing wansui*, edited by Tsai Ming-liang, et al., 192–207. Taipei: Wanxiang Tushu, 1994.

Chen, Huisi. "Dianjianju chajin Cai Mingliang *Hei yanquan*-zuiming: fanying choulou mian buli lüyounian." *Duli Xinwen Zaixian*. February 26, 2007. http://www.merdekareview.com/news.php?n=3537.

Chen Shuzhe, "Exception or Useless? The enigma to the costume of the first lady", 2013, see Forbes, accessed 10th May, 2013. http://www.forbeschina.com/review/201303/0024525.shtml

Cheng, Chung-Ying. "On the Metaphysical Significance of *Ti* (Body-Embodiment) in Chinese Philosophy: *Benti* (Origin-Substance) and *Ti-Yong* (Substance and Function)." *Journal of Chinese Philosophy* 29, no. 2 (June 2002): 14561.

Chi, Robert. "*The Red Detachment of Women*: Resenting, Regendering, Remembering." In *Chinese Films in Focus II*, edited by Chris Berry, 152–59. London: British Film Institute, 2003.

Chiu, Kuei-fen. "Afterword: Documentary Filmmaking as Ethical Production of Truth." *Concentric: Literary and Cultural Studies* 39, no. 1 (March 2013): 203–20.

Chivers, Meredith, Gerulf Riefer, Elizabeth Latty, and Michael Bailey. "A Sex Difference in the Specificity of Sexual Arousal." *Psychological Science* 4, no. 11 (2004): 736–44.

Chodorow, Nancy. *The Reproduction of Mothering: Psychoanalysis and the Sociology of Gender*. Los Angeles: University of California Press, 1978.

Chow, Rey. *Primitive Passions: Visuality, Sexuality, Ethnography, and Contemporary Chinese Cinema*. New York: Columbia University Press, 1995.

———. *Sentimental Fabulations, Contemporary Chinese Films: Attachment in the Age of Global Visibility*. New York: Columbia University Press, 2007.

Chu, Wei-cheng Raymond. "Taiwan tongzhi yundong de houzhimin sikao: lun 'xianshen' wenti." In *Pipan de xing zhengzhi: Tai she xingbie yu tongzhi duben*, edited by Chu Wei-cheng Raymond, 191–213. Taipei: Taiwan shehui yanjiu zazhishe, 2008.

———. "Tongzhi. Taiwan: Xing gongmin, guozu jiangou huo gongmin shehui." *Nüxue Xuezhi* 15 (May 2003): 115–51.

Chun, Wendy. *Control and Freedom: Power and Paranoia in The Age of Fiber Optics*. Cambridge, MA: MIT Press, 2006.

Colman, Felicity. "Affect." In *The Deleuze Dictionary*, edited by Adrian Parr, 11–13. Rev. ed. Edinburgh: Edinburgh University Press, 2010.

Dai, Jinhua. *Wuzhong fengjing: Zhongguo dianying wenhua, 1978–1998*. Beijing: Beijing da xue chu ban she, 2006.

Dai, Yan. *Meili Yu Aichon-Yige Zhenshide Ruan Lingyu* (Beauty and sadness - the real Ruan Lingyu). Beijing: Dongfang Chubanshe, 2005.

Daniels, Timothy P. *Building Cultural Nationalism in Malaysia*. New York: Routledge, 2005.

de Certeau, Michel. *The Practice of Everyday Life*, translated by Steven Rendall. Berkeley: University of California Press, 1984.

de Luca, Tiago. "Sensory Everyday: Space, Materiality and the Body in the Films of Tsai Ming-liang." *Journal of Chinese Cinemas* 5, no. 2 (2011): 157–79.

Deleuze, Gilles. *Cinema 2: The Time-Image*. Translated by Hugh Tomlinson and Robert Galeta. Minneapolis: University of Minnesota Press, 2010.

———. *Spinoza: Practical Philosophy*. Translated by R. Hurley. San Francisco: City Lights Books, 1988.

———. *What Is Philosophy?* Edited by Lawrence D. Kritzman. Translated by Hugh Tomlinson and Graham Burchell. New York: Columbia University Press, 1994.

Deleuze, Gilles, and Guattari, Félix. *A Thousand Plateaus: Capitalism and Schizophrenia*. Translated by Brian Massumi. Minneapolis: University of Minnesota Press, 2004.

Devraux, Alain. "Mirror Man." *Daily Tiger* 5. 39th International Film Festival Rotterdam. February 1, 2010.

Doane, Mary Ann. *The Emergence of Cinematic Time*. Cambridge, MA: Harvard University Press, 2002.

Dyer, Richard. *Stars*. London: British Film Institute, 1998.

Fainaru, Dan. "Useless (*Wu Yong*)." *ScreenDaily*. 2007. http://www.screendaily.com/useless-wu-yong/4034477.article.

Fatemi, Sasan. "Music, Festivity, and Gender in Iran from the Qajar to the Early Pahlavi Period." *Iranian Studies* 3, no. 3 (2005): 399–416.

Felman, Shoshana. "Film as Witness: Claude Lanzmann's *Shoah*." In *Holocaust Remembrance: The Shapes of Memory*, edited by Geoffrey Hartman. Oxford: Blackwell, 1994.

Finnane, Antonia. *Changing Clothes in China: Fashion, History, Nation*. New York: Columbia University Press, 2013.

Foucault, Michel. *Discipline and Punish: The Birth of the Prison*. Translated by Alan Sheridan. New York: Vintage, 1977.

Fu, Poshek, and Lane J. Harris. *China Forever: The Shaw Brothers and Diasporic Cinema. Popular Culture and Politics in Asia Pacific*. Urbana: University of Illinois Press, 2008.

Gilbert, Jeremy. "Signifying Nothing: 'Culture,' 'Discourse' and the Sociality of Affect." *Culture Machine* 6 (2004).

Gonzalez-Arnal, Stella, Gill Jagger, and Kathleen Lennon, eds. *Embodied Selves*. New York: Palgrave Macmillan, 2012.

Gregg, Melissa, and Gregory J. Seigworth. "An Inventory of Shimmers." In *The Affect Theory Reader*, edited by Melissa Gregg and Gregory J. Seigworth, 1–25. Durham, NC: Duke University Press, 2010.

Grosz, E.A. *Volatile Bodies: Toward a Corporeal Feminism*. Bloomington: Indiana University Press, 1994.

Grosz, Elizabeth. *Becomings: Exploration in Time, Memory and Futures*. Ithaca, NY: Cornell University Press, 1999.

———. *Volatile Bodies: Toward a Corporeal Feminism*. Bloomington: Indiana University Press, 1994.

Gullett, Margaret Morganroth. "Postmaternity as a Revolutionary Feminist Concept." *Feminist Studies* 28, no. 3 (2002): 553–72.

Halberstam, Jack "Judith." *The Queer Art of Failure*. Durham, NC: Duke University Press, 2011.

Harris, Kristine. "*The Goddess*: Fallen Woman of Shanghai." In *Chinese Films in Focus II*, edited by Chris Berry, 128–36. Basingstoke, England: Palgrave Macmillan, 2008.

———. "The *New Woman* Incident: Cinema, Scandal, and Spectacle in 1935 Shanghai." In *Transnational Chinese Cinemas: Identity, Nationhood, Gender*, edited by Sheldon Hsiao-peng Lu, 277–302. Honolulu: University of Hawaii Press, 1997.

Hart, Kylo-Patrick R. *Queer Males in Contemporary Cinema: Becoming Visible*. Lanham, MD: Scarecrow Press, 2013.

Hay, John. "The Body Invisible in Chinese Art?" In *Body, Subject, and Power in China*, edited by Angela Zito and Tani E. Barlow, 42–77. Chicago: University of Chicago Press, 1994.

Hefner, Carl J. "Ludruk Folk Theater of East Java: Toward a Theory of Symbolic Action." PhD diss., University of Hawaii, 1994.

Heng, Geraldine, and Janadas Devan. "State Fatherhood: The Politics of Nationalism, Sexuality, and Race in Singapore." In *Nationalisms and Sexualities*, edited by Andrew Parker, Mary Russo, Doris Sommer, and Patricia Yaegar, 343–64. New York: Routledge, 1992.

Higson, Andrew. "The Limiting Imagination of National Cinema." In *Cinema and Nation*, edited by M. Hjort and S. MacKenzie. New York: Routledge, 2000.

Hjort, Mette. "Ruan Lingyu: Reflections on an Individual Performance Style." In *Chinese Film Stars*, edited by Mary ed Farquhar, introd and Yingjin Zhang. Routledge Contemporary China Series (Routledge Contemporary China Series): 51, 32-49. London, England: Routledge, 2010.

Ho, Chi Sam. "Transgender Representations." Master's thesis, Department of Communications and New Media, National University of Singapore, 2010.

Hodgson, Marshall G. S. *The Venture of Islam: Conscience and History in a World Civilization*. Chicago: University of Chicago Press, 1974.

Honohan, Iseult.. Metaphors of solidarity. In *Political language and metaphor. Interpreting and changing the world*, edited by T. Carver and J. Pikalo. New York: Routledge, 2008

Hu, Brian. "Star Discourse and the Cosmopolitan Chinese: Linda Lin Dai Takes on the World." *Journal of Chinese Cinemas* 4, no. 3 (2010): 183–209.

Ince, Kate. "Queering the Family? Fantasy and the Performance of Sexuality and Gay Relations in French Cinema, 1995–2000." *Studies in French Cinema* 2, no. 2 (2002): 90–97.

Jacobs, Katrien. *People's Pornography: Sex and Surveillance on the Chinese Internet*. Bristol, UK: Intellect Books, 2011.

Jiang, Huifen. "Cai Mingliang, Li Kangsheng 'tongzhi qing' gaobai." *TVBS*. April 17, 2003. http://www.tvbs.com.tw/news/news_list.asp?no=jcw62020030417021012.

Jiang siyuan, "The female actresses who follow the first lady and wear Exception", 2013. see China.com.cn. accessed 10th May, 2013. http://ent.china.com.cn/2013-03/27/content_28370607.htm

Johnson, Galen A. "Ontology and Painting 'Eye and Mind'" translated by Michael B. Smith. In *The Merleau-Ponty Aesthetics Reader: Philosophy and Painting*, edited by Maurice Merleau-Ponty, Galen A. Johnson, and Michael B. Smith, 35–56. Evanston, IL: Northwestern University Press, 1993.

Johnston, Cristina. "Representations of Homosexuality in 1990s Mainstream French Cinema." *Studies in French Cinema* 2, no. 1 (2002): 23–31.

Kar, L., F. Bren, and S. Ho. *Hong Kong Cinema: A Cross-Cultural View*. Lanham: Scarecrow Press, 2004.

Karlyn, Kathleen Rowe. *Unruly Girls, Unrepentant Mothers: Redefining Feminism On Screen*. Austin: University of Texas Press, 2011.

Keenan, Dennis King. *On the Question of Sacrifice*. Bloomington: Indiana University Press, 2005.

Khoo, Olivia. "Love in Ruins: Spectral Bodies in Wong Kar-wai's *In the Mood for Love*." In *Embodied Modernities: Corporeality, Representation, and Chinese Cultures*, edited by Fran Martin and Larissa Heinrich, 235–52. Honolulu: University of Hawai'i Press, 2006.

Kuhn, Annette. *Cinema, Censorship, and Sexuality*. New York: Routledge, 1990.

Kim, Jihoon. "Between Auditorium and Gallery: Perception in Apichatpong Weerasethakul's Films and Installations." In *Global Art Cinema: New Theories and Histories*, edited by Rosalind Galt and Karl Schoonover, 125-41. Oxford: Oxford University Press, 2010.

Kinder, Marsha. *Blood Cinema: The Reconstruction of National Identity in Spain*. Berkeley: University of California Press, 1993.

Kristeva, Julia. "Stabat Mater." In *Tales of Love*, translated by Leon S. Roudiez, 234–63. New York: Columbia University Press, 1997.

Kuhn, Annette. *Cinema, Censorship, and Sexuality*. New York: Routledge, 1990.

Lam, Edward. "Xiao tuanyuan-guancha *Heliu*." *Mingpao*. April 27, 1997.

Law, Kar, Frank Bren, and Sam Ho. *Hong Kong Cinema: A Cross-Cultural View*. Lanham, MD: Scarecrow Press, 2004.

Lawrence, Michael. "Lee Kang-sheng: Non-professional star." In *Chinese Film Stars*, edited by Mary Farquhar and Ying-jin Zhang, 151–62. New York: Routledge, 2010.

Lee, Hock Guan. "Affirmative Action in Malaysia." In *Southeast Asian Affairs*, edited by Chin Kin Wah and Daljit Singh, 211–28. Singapore: Institute of Southeast Asian Studies, 2005.

Lee, Kang-sheng. Recorded interview by Wai-siam Hee in Taipei, June 21, 2010.

Lee, Terence. *The Media, Cultural Control and the Government in Singapore*. New York: Routledge, 2010.

Lee, Yuen Beng. *Transnational Malaysian Cinema (2000 - 2012). New Forms and Genres in post-2000 Malaysian Cinema*. Melbourne: The University of Melbourne, 2012.

Leung, Helen Hok-Sze. "Trans on Screen." In *Transgender China*, edited by Howard Chiang. New York: Palgrave, Macmillan, 2012.

Lim, Song Hwee. *Celluloid Comrades: Representations of Male Homosexuality in Contemporary Chinese Cinemas*. Honolulu: University of Hawai'i Press, 2006.

———. "Positioning Auteur Theory in Chinese Cinemas Studies: Intratextuality, Intertextuality and Paratextuality in the Films of Tsai Ming-liang." *Journal of Chinese Cinemas* 1, no. 3 (2007): 229–36.

Lim, Song Hwee, and Hee Wai-siam. "Ni bixu xiangxin dianying you yige zuozhe: zai lutedan fangtan Cai Mingliang." *Dianying Xinshang* 147 (April-June 2011): 69–76.

———. "You Must Believe There Is an Author behind Every Film: An Interview with Tsai Ming-liang." *Journal of Chinese Cinemas* 5, no. 2 (2011): 181–91.

Lin, Meilian. "Porn's Peephole." *Global Times*, October 30, 2011. http://www.globaltimes.cn/NEWS/tabid/99/ID/681579/Porns-peephole.aspx.

Lionnet, Françoise, and Shu-mei Shih. *The Creolization of Theory*. Durham, NC: Duke University Press, 2011.

Liu, Jen-peng, and Ding Nai-fei. "Reticent Poetics, Queer Politics." In *The Inter-Asia Cultural Studies Reader*, edited by Kuan-Hsing Chen, 395–424. New York: Routledge, 2007.

Louie, Kam. Introduction to *Hong Kong Culture: Word and Image*, edited by Kam Louie, 1–8. Hong Kong: Hong Kong University Press, 2010.

Lovatt, Philippa. "The Spectral Soundscapes of Postsocialist China in the Films of Jia Zhangke." *Screen* 53, no. 4 (Winter 2012): 418–35.

Lu, Xun. *Lu Xun Quanji*. Vol. 4. Beijing: Renmin wenxue chubanshe, 1973.

Lumby, Catherine, Kath Albury, and Alan McKee, eds. *The Porn Report*. Melbourne: Melbourne University Publishing, 2008.

Ma, Jean. "Photography's Absent Times." In *Melancholy Drift: Marking Time in Chinese Cinema*, 51–71. Hong Kong: Hong Kong University Press, 2002.

Marchetti, Gina. "*Transamerica*: Queer Cinema in the Middle of the Road." In *Hit the Road, Jack: Essays on the Culture of the American Road*, edited by Gordon E. Slethaug and Stacilee Ford, 198–213. Quebec: McGill-Queen's University Press, 2012.

Marcuse, Herbert. "Beyond One-Dimensional Man." *Towards a Critical Theory of Society: Collected Papers of Herbert Marcuse*. Vol. 2. London: Routledge, 2001.

———. *Eros and Civilization*. New York: Vintage Books, 1955.

———. *An Essay on Liberation*. Boston: Beacon Press, 1969.

———. *One-Dimensional Man*. Boston: Beacon Press, 1969.

———. "Beyond One-Dimensional Man." *Towards a Critical Theory of Society: Collected Papers of Herbert Marcuse*, vol. 2. London: Routledge, 2001

Marks, Laura. "Signs of the Time: Deleuze, Peirce and the Documentary Image." In *The Brain Is the Screen: Gilles Deleuze's Cinematic Philosophy*, edited by Gregory Flaxman, 193–214. Minneapolis: University of Minnesota Press, 2000.

———. *The Skin of the Film: Intercultural Cinema, Embodiment, and the Senses*. Durham, NC: Duke University Press, 2000.

———. *Touch: Sensuous Theory and Multisensory Media*. Minneapolis: University of Minnesota Press, 2002.

Martin, Fran. "The European Undead: Tsai Ming-liang's Temporal Dysphoria." *Senses of Cinema* 27 (July 2003). http://www.sensesofcinema.com/2003/feature-articles/tsai_european_undead/.

Martin, Fran, and Larissa Heinrich, "Introduction to Part 1: Thresholds of Modernity." In *Embodied Modernities: Corporeality, Representation, and Chinese Cultures*, edited by Fran Martin and Larissa Heinrich, 1–30. Honolulu: University of Hawai'i Press, 2006.

———, eds. *Embodied Modernities: Corporeality, Representation, and Chinese Cultures*. Honolulu: University of Hawai'i Press, 2006.

Massumi, Brian. *Parables for the Virtual: Movement, Affect, Sensation*. Durham, NC: Duke University Press, 2002.

Mazzarella, William. "Affect: What Is It Good For?" In *Enchantments of Modernity: Empire, Nation, Globalization*, edited by Saurabh Dube, 291–309. London: Routledge, 2009.

McGrath, Jason. *Postsocialist Modernity: Chinese Cinema, Literature and Criticism in the Market Age*. Stanford, CA: Stanford University Press, 2008.

Ni, Ba. "Kelamayi daoyan xu xin fangtan." *Fanhall.com*. 2010. http://gsz2006.i.sohu.com/blog/view/154025600.htm.

Olkowski, Dorothea. *Gilles Deleuze and the Ruin of Representation*. Berkeley: University of California Press, 1999.

Oshima, Nagisa. "Sex, Cinema, and the Four-and-a-Half Mat Room." In *Cinema, Censorship, and the State: The Writings of Nagisa Oshima*, edited by Annette Michaelson, translated by Dawn Lawson, 249–50. Cambridge, MA: MIT Press, 1993.

Paasonen, Susanna. *Carnal Resonance: Affect and Online Pornography*. Cambridge, MA: MIT Press, 2011.

Pang, Laikwan. *Building a New China in Cinema: The Chinese Left-Wing Cinema Movement, 1932–1937*. Lanham, MD: Rowman & Littlefield Publishers, 2002.

———. "The Making of a National Cinema: Shanghai Films of the 1930s." In *The Chinese Cinema Book*, edited by Song Hwee Lim, Julian Ward, Rey Chow, and Zou Yijie, 56–64. Basingstoke, England: Palgrave Macmillan, 2011.

Pavsek, Christopher. *The Utopia of Film: Cinema and Its Futures in Godard, Kluge, and Tahimik*. New York: Columbia University Press. 2013.

Peacock, James. *Rites of Modernization: Symbolic and Social Aspects of Indonesian Proletarian Drama*. Chicago: University of Chicago Press, 1987.

Peng, Hsiao-yen. "*Auteurism* and Taiwan New Cinema." *Journal of Theater Studies* 9 (January 2012): 125–48.

———. "*Haijiao tianya*: yiwai de chenggong? Huigu Taiwan xin dianying." *Dianying xinshang xuekan* 142 (January-March 2010): 124–36.

Pettys, Rebecca Ansary. "The Ta'zieh: Ritual Enactment of Persian Renewal." *Theatre Journal* 33, no. 3 (1981): 341–54.

Picard, Andréa. "Programmer's Note." TIFF. 2012. http://tiff.net/filmsandschedules/tiff/2012/walker.

Proctor-Xu, Jami. "Sites of Transformation: The Body and Ruins in Zhang Yang's *Shower*." In *Embodied Modernities: Corporeality, Representation, and Chinese Cultures*, edited by Fran Martin and Larissa Heinrich, 162–76. Honolulu: University of Hawai'i Press, 2006.

Pugsley, Peter C. "Postsocialist Modernity: Chinese Cinema, Literature and Criticism in the Market Age." *Continuum: Journal of Media & Cultural Studies* 24, no. 2 (2010): 325–27.

Raffles, Hugh. "Intimate Knowledge." *International Social Science Journal* 54, no. 3 (2002): 325–35.

Rancière, Jacques. *The Politics of Aesthetics*. New York: Continuum, 2006.

Reynaud, Bérénice. "Translating the Unspeakable: On-Screen and Off-Screen Voices in Wu Wenguang's Documentary Work." In *The New Chinese Documentary Film Movement: For the Public Record*, edited by Chris Berry, Lu Xinyu, and Lisa Rofel, 157–76. Hong Kong: Hong Kong University Press, 2010.

Rich, Adrienne. *Of Woman Born: Motherhood as Experience and Institution*. New York: W. W. Norton, 1976.

Ritzer, George, ed. *The Blackwell Companion to Globalization*. Hoboken, NJ: John Wiley & Sons, 2008.

Riviere, Daniele. *Dingwei: Yu Cai Mingliang de fangtan*. Taipei: Yuanliu, 2001.

Robinson, Jennifer. "Dying To Tell: Lesbian Practice and Love Suicide in Early Twentieth Century Japan." In *Queer Diasporas*, edited by Cindy Patton and Benigno Sanchez-Eppler, 38–70. Durham, NC: Duke University Press, 2000.

Rofel, Lisa. *Desiring China: Experiments in Neoliberalism, Sexuality, and Public Culture*. Durham, NC: Duke University Press, 2007.

Ruddick, Susan. "The Politics of Affect: Spinoza in the Works of Negri and Deleuze." *Theory, Culture and Society* 27, no. 4 (2010): 21–45.

Sarris, Andrew. *The American Cinema: Directors and Directions, 1929-1968*. Chicago: The University of Chicago Press, 1985.

Schaefer, William. "Shadow Photographs, Ruins, and Shanghai's Projected Past." *PMLA* 122, no. 1 (2007): 124–34.

Seow, Betty, and Beng Guat. Interview (Accession 1048/17), Oral History Center, National Archives Singapore.

Shih, Shu-Mei. "Against Diaspora: The Sinophone as Places of Cultural Production." *Global Chinese Literature*, edited by Jing Tsu and David Der-wei Wang, 29–48. Leiden, Netherlands: Brill, 2010.

———. Introduction to *Sinophone Studies: A Critical Reader*, edited by Shu-mei Shih, Brian Bernards, and Chien-hsin Tsai, 1–16. New York: Columbia University Press, 2013.

———. *Visuality and Identity: Sinophone Articulations across the Pacific*. Berkeley: University of California Press, 2007.

Shu, Liu. "Hongyan Boming Di Meili Jiaren: Lin Dai." *Dazhong dianying*, 48–53. June 2004.

Si, Han. "China's Richest 20 Cities." *Sohu.com*. 2012. http://business.sohu.com/20120327/n339020485.shtml.

Simmel, George, ed. *The Philosophy of Money*. 3rd ed. Translated by Tom Bottomore and David Frisby. New York: Routledge, 2004.

Sloterdijk, Peter. *Critique of Cynical Reason*. Trans. Michael Eldred. Minneapolis: University of Minnesota Press, 1987.

Smith, Anthony D. *National identity*. London; New York: Penguin, 1991.

Smith, Clarissa, Feona Attwood, and Martin Barker. "UK Porn Research Preliminary Statistics." 2012. http://www.pornresearch.org/Firstsummaryforwebsite.pdf.

Smith, Kathie. "VIFF 2012 Review: BEAUTIFUL 2012 Is a Rare, Successful Omnibus." *Twitch*. 2012. http://twitchfilm.com/2012/10/viff-2012-review-beautiful-2012-is-a-rare-successful-omnibus.html.

Snyder, Sharon L., and David T. Mitchell. *Cultural Locations of Disability*. Chicago: University of Chicago Press, 2006.

Sobchack, Vivian. *Carnal Thoughts: Embodiment and Moving Image Culture*. Berkeley: University of California Press, 2004.

Stam, Robert. *Film Theory: An Introduction*. Malden: Blackwell Publishing, 2000.

Stryker, Susan, Paisley Currah, and Lisa Jean Moore. Introduction to *WSQ: Women's Studies Quarterly* 36, nos. 3 & 4 (2008).

Stuttgen, Tim, ed. *Post/Porn/Politics: Queer and Feminist Perspectives on the Politics of Porn Performance and Sex Work as Cultural Production*. Berlin: B_books, 2010.

Su, Tong. *Raise the Red Lantern: Three Novellas*. Translated by Michael Duke. London: Scribner, 1993.

Tan, E. K. "Transcending Multiracialism: Kuo Pao Kun's Multilingual Play *Mama Looking for Her Cat* and the Concept of *Open Culture*." In *Sinophone Studies: A Critical Reader*, edited by Shu-mei Shih, Brian Bernards, and Chien-hsin Tsai, 315–25. New York: Columbia University Press, 2013.

Tan, Kenneth Paul. *Cinema and Television in Singapore: Resistance in One Dimension*. Leiden, Netherlands: Brill, 2008.

Tan, Sooi Beng. *Bangsawan: A Social and Stylistic History of Popular Malay Opera*. Oxford: Oxford University Press, 1993.

Teo, Stephen. "China with an Accent: Interview with Jia Zhangke, Director of *Platform*." *Senses of Cinema*. 2001. http://sensesofcinema.com/2001/feature-articles/zhangke_interview/.

Tian, Si. "Wenxue Cai Mingliang." *Chao Foon* 489 (December 2002): 30–37.

Tsai, Ming-Liang. "Director's Notes," in *I Don't Want to Sleep Alone* (DVD), Taipei: Homegreen Films, 2006.

Tsai, Ming-liang. "Wo meiyou wangzhi." *Chao Foon* 489 (December 2002): 52–53.

Tsoi, Wing Foo. "Developmental Profile of 200 Male and 100 Female Transsexuals in Singapore." *Archives of Sexual Behavior* 19, no. 6 (1990): 595–605.

———. "The Prevalence of Transsexualism in Singapore." *Acta Psychiatrica Scandinavica* 78 (1988): 501–4.

Vandenberghe, Frédéric. "Deleuzian Capitalism." *Philosophy and Social Criticism* 34, no. 8 (2008): 877–903.

Voros, Florian. Forthcoming, "Domesticated porn: Gendered Embodiment in Audience Reception Practices of Pornography. In E. Biasin, G. Maina and F. Zecca, *Porn after Porn: Contemporary Alternative Pornographies* Milano, Mimesis, 2014.

Walker, Rebecca. *Baby Love: Choosing Motherhood after a Lifetime of Ambivalence.* New York: Penguin Books, 2007.

Wang, Haowei. "Buyao jiao chu yaokongqi: tongzhi yao you 'xianshen' zizhuquan." Taiwan *Saodong* 3 (January 1997): 52–58.

Wang, Molin. "Bei bianchu de jiashen: Cai Mingliang dianying zhong de fu yu zi." *Dianying Xinshang* (March 2002): 71–75.

Wen, Tianxiang. *Guangying dingge: Cai Mingliang de xinling changyu.* Taipei: Hengxing guoji wenhua, 2002.

Williams, Linda. *Hard Core: Power, Pleasure, and the "Frenzy of the Visible."* Berkeley: University of California Press, 1989.

Wong, Kim Hoh. "When Papa became Mama." *The Straits Times.* September 8, 2008.

Wong, Kokkeong. *Media and Culture in Singapore: A Theory of Controlled Commodification.* Cresskill, NJ: Hampton Press, 2001.

Wong, Nicholas Y. B. "Film Reviews: 'The Hand.'" *Scope* 30, no. 4 (August 2013).

Xu Jinrong, Ye Wanru, and Zheng Weibo. "Cai Mingliang ci sheng bu hun, qing xi Li Kangsheng." Taiwan *Pingguo ribao.* March 25, 2007. http://ent.appledaily.com.tw/enews/article/entertainment/20070325/3343378/.

Yang, Jeff, and Art Black. *Once Upon a Time in China: A Guide to Hong Kong, Chinese, and Taiwanese Cinema.* New York: Atria Books, 2003.

Yao, Yiwei. "Yibu meiyou jia de dianying: wo kan *Aiqing wansui.*" *Sinchew Daily* (Malaysia). July 10, 1994.

Yau, Ching. "Porn Power: Sexual and Gender Politics." In *As Normal as Possible: Negotiating Sexuality and Gender in Mainland China and Hong Kong*, edited by Yau Ching, 113–30. Hong Kong: Hong Kong University Press, 2010.

Yau, Herman, and Anthony Wong. *Ebola Syndrome*. Audio commentary to Discotek Media (US) DVD release, 2007.

Yeh, Emilie Yueh-Yu, and Darrell William Davis. *Taiwan Film Directors: A Treasure Island*. New York: Columbia University Press, 2005.

Zhang, Caihong. *Shenti Zhengzhi: Bainian Zhongguo Dianying Nü Mingxing Yanjiu*. Beijing: Zhongguo guangbo dianshichubanshe, 2011.

Zhang, Yingjin. *Chinese National Cinema*. New York: Taylor & Francis, 2004.

Zhang, Zhen. *An Amorous History of the Silver Screen: Shanghai Cinema, 1896–1937*. Chicago: University of Chicago Press, 2005.

———. "The Negating Subject in Progressive Time: Jia Zhangke's Xiao Wu." *International Journal of Humanities and Social Science* 1, no. 18 (2006): 162–72.

———, ed. *The Urban Generation: Chinese Cinema and Society at the Turn of the Twenty-First Century*. Durham, NC: Duke University Press, 2007.

Zhou, Huashan. *Houzhimin tongzhi*. Hong Kong: Xianggang tongzhi yanjiushe, 1997.

Zito, Angela, and Tani E. Barlow. "Introduction: Body, Subject, and Power in China." In *Body, Subject, and Power in China*, edited by Angela Zito and Tani E. Barlow, 1–22. Chicago: University of Chicago Press, 1994.

Filmography

3D Sex and Zen: Extreme Ecstasy (*3D rouputuan zh ji le baojian*), Christopher Sun Lap Key, Hong Kong, 2011.

Accident, The (*Xinyuan yima*), Julian Lee, Hong Kong, 1999.

Accused, The, Jonathan Kaplan, Canada, 1988.

All Corners of the World, TV drama, Tsai Ming-liang, Taiwan, 1989.

As I Lay Dying, Ho Yuhang, Malaysia, 2006.

As Tears Go By (*Wong gok ka moon*), Wong Kar-wai, Hong Kong, 1988.

At the End of Daybreak (*Sham Moh*), Ho Yuhang, South Korea/Malaysia/Hong Kong, 2009.

Aunty Lucy Slam Dunk (*Aunty Lucy Ye guan lan*), Gerald Lee, Singapore, 2009.

Be with Me, Eric Khoo, Singapore, 2005.

Beasts, The (*Shangou*), Dennis Wu, Hong Kong, 1980.

Beautiful Washing Machine, The (*Meilide xiyiji*), James Lee, Malaysia, 2004.

Boys, (*Xiao Hai*) Tsai Ming-Liang, Taiwan, 1991.*Big Heat, The* (*Chengshi tejing*), Johnnie To Andrew Kam, and Tsui Hark, Hong Kong, 1988.

Blonde Fury, The (*Shijie dashai*), Hoi Mang, Hong Kong, 1989.

Brother of Darkness (*Titian xingdao zhi shaxiong*), Billy Tang, Hong Kong, 1994.

Call If You Need Me, James Lee, Malaysia, 2009.

Cinderella and Her Little Angels (*Yun chang yan hou*), Tang Huang, Hong Kong, 1959.

Company of Mushrooms (*Mo gu xiong di men*), Tan Chui Mui, Malaysia, 2006.

Crazy (*Siji sharenkuang 2: Nanfeng*), John Hau, Hong Kong, 1999.

Daughter of Darkness (*Miemen canan zhi niesha*), Lai Kai-Ming, Hong Kong, 1993.

Daughter of Darkness 2 (*Miemen canan 2 zhi jiezhong*), Lai Kang-Ming, Hong Kong, 1994.

Day without a Policeman, A (*Moujing shifen*), Johnny Lee, Hong Kong, 1993.

Daze Raper (*Mijian fan*), Wilson Yip, Hong Kong, 1995.

Don't Play with Fire (*Diyi leixing weixian*), Tsui Hark, Hong Kong, 1980.

Dragon Inn a.k.a. *New Dragon Inn* (*Xin Longmen kezhan*), Raymond Lee, Tsui Hark, and Ching Sui-Tung, Hong Kong, 1992.

Dreaming the Reality (*Mengxing xue weiting*), Wong Chun-Yeung, Hong Kong, 1991.

Dr. Lamb (*Gaoyang yisheng*), Danny Lee, Hing Sing Billy Tang, Hong Kong, 1992.

East Is Red, The (*Dongfang bubai: Fengyun zaiqi*), Tsui Hark, Hong Kong, 1993.

East Palace, West Palace (*Donggong xigong*), Zhang Yuan, China, 1996.

Ebola Syndrome (*Yibola bingdu*), Herman Yau, Hong Kong, 1996.

Elephant and the Sea, The, Woo Ming Jin, Malaysia/Netherlands, 2003.

Enter the Dragon, Robert Clouse, Hong Kong/USA, 1973.

Eros, anthology with three short segments directed by Wong Kar-wai ("The Hand"), Steven Soderbergh ("Equilibrium"), and Michelangelo Antonioni ("The Dangerous Thread of Things"), Hong Kong/USA/Italy, 2004.

Eternal Evil of Asia, The (*Nanyang shida xieshu*), Man Kei Chin, Hong Kong, 1995.

Face (*Visage*), Tsai Ming-liang, France/Taiwan/Belgium/Netherlands, 2009.

Farewell, China (*Ai zai biexiang de jijie*), Clara Law, Hong Kong, 1990.

Flirting Scholar (*Tang bohu dian qiuxiang*), Lik-Chi Lee, Hong Kong/China, 1993.

Flower in the Pocket (*Koudai li de hua*), Liew Seng Tat, Malaysia, 2007.

Fuck Cinema (*Cao ta ma de dianying*), Wu Wenguang, China, 2006.

Give Me a Home, TV drama, Tsai Ming-liang, Taiwan, 1991.

Goddess, The (*Shennü*), Wu Yonggang, China, 1934.

Golden Swallow (*Jin yanzi*), Chang Chen, Hong Kong, 1968.

Goodbye, Dragon Inn (Bu San), Tsai Ming-liang, Taiwan, 2003.

Grandmasters, The (*Yi dai zong shi*), Wong Kar-wai, Hong Kong/China, 2013.

Great Day (*Tiantian Haotian*), Chiu Keng Guan, Malaysia, 2011.

Happy Times (*Xinfu Shiguang*), Zhang Yimou, China, 2000.

He and She (*Jie mei qing shen*), Lawrence Cheng, Hong Kong, 1994.

He's a Woman, She's a Man (*Gam chi yuk sip*), Peter Chan and Chi Lee, Hong Kong, 1994.

Her Fatal Ways (*Biaojie, ni haoye!*), Alfred Cheung, Hong Kong, 1991.

Her Vengeance (*Xue meigui*), Ngai Choi Lam, Hong Kong, 1988.

Hero Dream (*Yinyao haoqing*), Lau Keung-Fu, Hong Kong, 1993.

Heroic Trio, The (*Dongfang sanxia*), Johnnie To, Hong Kong, 1992.

Homecoming (*Xiaozhe huijia*), Lee Thean-Jeen, Singapore, 2011.

I Don't Want to Sleep Alone (*Hei yan quan*), Tsai Ming-liang, Malyasia/Taiwan, 2006.

Ice Kacang Puppy Love (*Chulian hongdou bing*), Ah Niu, Malaysia, 2010.

If It's Not Now, Then When? (*Ruguo haiyou mingtian*), James Lee, Malaysia, 2012.
Illegal Immigrant, The (*Feifa yimin*), Mabel Cheung, Hong Kong, 1985.
In the Mood for Love (*Fa yeung nin wa*), Wong Kar-wai, Hong Kong/France, 2000.
Ju Dou, Yang Fenliang and Zhang Yimou, China/Japan, 1990.
Judgment Day (*Shijie mori*), Ong Kuo Sin, Singapore, 2013.
Karamay (*Kelamayi*), Xu Xin, China, 2010.
Kinta 1881, (Jinda), C. L. Hor, Malaysia, 2008.
Last Temptation of Christ, The, Martin Scorcese, USA/Canada, 1988.
Love Conquers All (El amor lo conquista todo?), Tan Chui Mui, Malaysia, 2006.
Love to Kill (*Nue zhi lian*), Chung Siu-hung and Kirk Wong, Hong Kong, 1993.
Love without End (*Bu liao qing*), Tao Qin, Hong Kong, 1961.
Lust, Caution (*Sejie*), Ang Lee, USA/China/Taiwan, 2007.
Mr. Vampire (*Jiangshi Xiansheng*), Ricky Lau, Hong Kong, 1985.
Naked Killer (*Chiluo gaoyang*), Clarence Ford, Hong Kong, 1992.
Nasi Lemak 2.0 (*Lasinima*), Namewee a.k.a. Wee Meng Chee, Malaysia, 2011.
New Women (*Xin nüxing*), Cai Chusheng, 1935.
Offense Storm (*Nutongdang xingfengbao*), Wong Ying-Git, Hong Kong, 1993.
Once Upon a Time in China IV (*Huang feihong zhi si: Wangzhe zhi feng*), Tsui Hark and Yuen Bun, Hong Kong, 1993.
Pantyhose Heroes (*Zhifen shuangxiong*), Sammo Hung Kam-bo, 1991.
Passion Unbounded (*Siji sharenkuang*), John Hau, Hong Kong, 1995.
Passionate Killing in the Dream (*Yunyu diliugan*), Kuo Chu Huang, Hong Kong, 1992.
Petaling Street Warriors (*Da yingxiong, xiao nanren*), James Lee, Malaysia, 2011.
Petition, Zhao Liang, China/Switzerland/UK/France/Belgium/Finland, 2011.
Pisau Cukur, Bernard Chauly, Malaysia, 2009.
Police Story Series (*Jingcha gushi xilie/Ging caat gusi hai lit*), Jackie Chan (*Police Story 1 & 2*), Stanley Tong (*Police Story 3 & 4*), Benny Chan Muk-Sing (*New Police Story*), Hong Kong, 1985.
Possessed, Bjarne Wong, Malaysia, 2006.
Power of Women at 40, The (女人40当自强), TV drama, Jack Neo, Television Corporation of Singapore Channel 8, 1997–2003.
Rain Dogs (*Tai yang yue*), Ho Yuhang, Singapore, 2006.

Raise the Red Lantern (*Dahong delong gaogao hua*), Zhang Yimou, Hong Kong/Taiwan/China, 1991.
Rebels of the Neon God (Ch'ing shaonien ne cha), Tsai Ming-liang, Taiwan, 1992.
Red Detachment of Women (*Hongse Niangzijun*), Xie Jin, China, 1964.
Red to kill (Roshan), Billy Tang, Hong Kong, 1994.
Righting Wrongs (*Zhifa xianfeng*), Corey Yuen, Hong Kong, 1986.
River, The (*He Liu*), Tsai Ming-liang, Taiwan, 1997.
Rock on Fire (*Jimi dangan zhi zhiming youhuo*), Lung Sang, Hong Kong, 1994.
Room to Let (*You Fangjian chuzu*), James Lee, Malaysia, 2002.
Run and Kill (*Wu syu*), Billy Tang, Hong Kong, 1993.
Sanctuary (*Wu*), Ho Yuhang, Malaysia, 2004.
"Sandra and Ralph: Audition," *Little Britain*, Series 1, Episode 7, written by Matt Lucas and David Williams, 2003.
Seed of Darkness (*Jing Ling*), Michael Chuah, Malaysia, 2006.
Sepet, Yasmin Ahmad, Malaysia, 2004.
Sex and Zen (*Yuputuan zhi touqing bao jian*), Michael Mak, 1991.
Shen Chunhua Life Show, "Zai Luofu Gong kanjian ta de lian: Cai Mingliang," TV broadcast, September 21, 2007. "http://v.youku.com/v_show/id_XMTIxOTUyOTY4.htm".
South of South, Tan Chui Mui, Malaysia, 2006.
Spider Lilies (*Ciqing*), Zero Chou, Taiwan, 2007.
Still Life (*Sanxia haoren*), Jia Zhangke, China, 2006.
Story of Qiu Jiu, The (*Qiujiu da guangsi*), Zhang Yimou, China, 1992.
Stray Dogs (Jiayou), Tsai Ming-liang, Taiwan/France, 2013.
Sweet Smell of Death, The (*Luoming geluofang*), Wilson Yip, Hong Kong, 1994.
Three Times (*Zuihaode shiguang*), Hou Hsiao-Hsien, Taiwan, 2005.
Tiger Woohoo (*Da rizi*), Chiu Keng Guan, Malaysia, 2010.
Transamerica, Duncan Tucker, USA, 2005.
Tree in Tanjung Malim, A, Chui Mui Tan, Singapore, 2004.
Trilogy of Lust (*Xuelian*), Julie Lee, Hong Kong, 1995.
Twist (*Zeiwong*), Danny Lee, Hong Kong, 1995.
Uncle Boonmee Who Can Recall His Past Lives (*Lung Bunmi Raluek Chat*), Apichatpong Weerasethakul, Thailand, 2010.
Under Police Protection (*Jinpai shijie*), Godfrey Ho, Hong Kong, 1990.

Untold Story, The (*Bat sin fan dim ji yan yuk cha siu bau*), Danny Lee and Herman Yau, Hong Kong, 1993.

Useless (*Wuyong*), Jia Zhangke, China, 2007.

Vive L'Amour (*Ai qing wan sui*), Tsai Ming-liang, Taiwan, 1994.

Walker (*Xingzhe*), Tsai Ming-liang, Hong Kong, 2012.

We Not Naughty (*Haizi bu hua*), Jack Neo, Singapore, 2012.

What Time Is It There? (*Ni Nabian Jidian*), Tsai Ming-liang, Taiwan, 2001.

Woman, Demon, Human (人鬼情 Rén Guǐ Qíng), Shuqin Huang, China, 1987

World, The (*Shijie*), Jia Zhangke, China, 2004.

The Wrath of Silence (*Chenmo de guniang*), Frankie Chan, Hong Kong, 1994.

Wuniu, Allen Fong, Hong Kong, 1990.

Xiao Wu (小武), Jia Zhangke, China, 1997.

Yellow Earth (*Huang tudi*), Chen Kaige, China, 1985.

Young Wife Violated before Her Husband's Eyes, A, Mao Kurata, Japan, 2010.

Zaijian Yulang (Goodbye, Fisherman)

Contributors

Brian Bergen-Aurand
Brian Bergen-Aurand teaches cinema and critical theory at Nanyang Technological University, Singapore, where he specializes in the relationship between film, ethics, and embodiment. He earned his PhD in comparative literature from the University of Maryland College Park in 2004 with a dissertation entitled "Seeing and the Seen: Post-Phenomenological Ethics and the Cinema." He is the author of *Film/Ethics: 1. Proper Names*. He is the editor of *Screen Bodies*, serves on the editorial board of *The New Review of Film and Television Studies*, and is the former sex and gender editor at *Clamor Magazine*. He has published essays on film, gender, sexuality, and embodiment in a number of journals.

Darren Byler
Darren Byler is a doctoral student in the Department of Anthropology at the University of Washington in Seattle. His research is focused on the phenomenology of urban life as expressed through the visual, material culture, and the built environment in Northwest China and Central Asia. His dissertation explores the ways in which differently positioned artists, filmmakers, musicians, and poets express cultural values and how these performances are in turn related to understandings of human ecology, ethics, and politics.

Andrew Grossman
Andrew Grossman is the editor of the anthology *Queer Asian Cinema: Shadows in the Shade*, a regular contributor to and editor of *Bright Lights Film Journal*, and a contributor to *The New Dictionary of the History of Ideas*. He has contributed book chapters to a number of critical anthologies, including *24 Frames: The Film of Korea and Japan* (Wallflower Press, 2004), *New Korean Cinema* (New York University Press, 2005), *Chinese Connections: Perspectives on Film, Identity, and Diaspora* (Temple University Press, 2009), *Film and Literary Modernism* (Cambridge Scholars, 2013), and *Asexualities: Feminist and Queer Perspectives* (Routledge, 2014).

Hee Wai-Siam

Hee Wai-Siam is assistant professor of Chinese at Nanyang Technological University, where he specializes in gender and sexuality in Chinese literature and culture. His articles on gender studies and Sinophone cinemas have appeared in *Queer Sinophone Cultures*, *Frontiers of Literary Studies in China*, *Film Appreciation Academic Journal*, *Taiwan: A Radical Quarterly in Social Studies*, and *Chung Wai Literary Quarterly*. He is currently working on a research project on the early history of Sinophone cinemas in Singapore and Malaya (1926–1965). He writes on gender and sexuality in Chinese literature and culture and is completing a history of sexuality in modern China.

Katrien Jacobs

Katrien Jacobs received a PhD in comparative literature and media from the University of Maryland at College Park. Her dissertation about 1960s/1970s performance art offers a unique blend of theoretical essays and video documentaries. She continues to work as a scholar and media artist who investigates the role of digital networks in people's experiences with the body, art, and sexuality. Her most recent book, *People's Pornography: Sex and Surveillance on the Chinese Internet* (Intellect Books, 2011), analyzes mainland China's immersion in new trends in sexual entertainment and DIY media. Her work can be found on www.libidot.org/blog.

Lee Yuen Beng

Lee Yuen Beng is currently senior lecturer at Universiti Sains Malaysia, Penang. His passion for academia began as a freelance English tutor. He then lectured full-time in a non-profit college before obtaining his PhD from Melbourne University. Currently, he explores issues of race and ethnicity, gender, and political economy in the cinemas, cultures, and medias of Malaysia and Asia.

Hongfei Liao

Hongfei Liao is a PhD candidate at The Amsterdam School for Cultural Analysis, University of Amsterdam, The Netherlands. Before that, he obtained his master's degree in film studies from Renmin University of China in 2011. His current research focuses on Chinese cinema and French film theory and philosophy. He is an online author and translator of http://cinephilia.net/, and he also translates and publishes books and articles on film and cultural theory. His Chinese translation of *Gilles Deleuze* (written by Claire Colebrook, Routledge, 2002) is coming out in the summer of 2014.

Mary Mazzilli

Mary Mazzilli is Research Associate at SOAS, University of London, UK. She was a fellow at Nanyang Technological University, Singapore. She specializes in Comparative Literature, Chinese cinema, the connection between urbanism and theater in Asia as well as gender and women's studies. Her first monograph Gao Xingjian's Post-Exile Plays: Transnationalism and Postdramatic Theatre is forthcoming with Methuen Drama-Bloomsbury Group.

Sim Jiaying

Sim Jiaying is a PhD candidate in the Film and Television Studies Department at the University of Glasgow, Scotland. She received both her bachelor of arts degree in English literature and master of arts degree in English from Nanyang Technological University, Singapore. Her research interests include the body and the senses with regard to embodied cinematic spectatorship, affect theory, film-philosophy, and transnational Chinese cinemas discourses.

Jun Zubillaga-Pow

Jun Zubillaga-Pow is a PhD candidate in music research at King's College London. His research interests include transnational cultures, queer sexualities, and psychoanalysis. Jun has published in *Sexualities*, *South East Asia Research*, and the *Journal of History and Cultures*, and is the coeditor of *Queer Singapore* (Hong Kong University Press, 2012) and *Singapore Soundscape* (National Library Board, 2014). He is currently working on the psychoanalysis of musical meaning, and the biographies of six pioneer composers in Singapore.

Index

3D Sex and Zen: Extreme Ecstasy 22, 225, 227, 231, 245, 247
accented cinema 182
Accident, The 216
A Day without a Policeman 211
affect 6-7, 11, 13, 19-20, 22, 30-31, 43, 51, 53, 55-57, 62, 64, 66-67, 71, 137-142, 146-148, 151-153, 157, 159-160, 162, 165-166, 169-170, 172, 179-180, 190, 225, 232, 235, 255, 262-263, 269-271, 279
Ah Niu 184-185, 250
Alfred Cheung 250
All Corners of the World 117-118, 121, 247
Allen Fong 208-209, 254
alterity 62, 71-75, 77, 81-82, 86
Althusser 180, 257
Andrew Kam 247
Ang Lee 43, 251
Anthony Appiah 125, 135
Antonioni 52, 56, 249
Apichatpong Weerasethakul 23, 25-26, 45, 253, 266
As I Lay Dying 185, 247
As Tears Go By 51, 247
A Tree in Tanjung Malim 191
At the End of Daybreak 188, 194, 247
Aunty Lucy Slam Dunk
auteur 5, 19, 113-116, 130-131, 134, 267
A Young Wife Violated before Her Husband's Eyes 22, 225, 227, 235
Bazin 116, 133, 258
Beasts, The 282
Beautiful Washing Machine, The 191, 247
Bernard Chauly 251
Be With Me 282

Big Heat, The 282
Billy Tang 247-248, 252
Biopolitics 6, 159
Blonde Fury, The 282
boys 39-40, 104, 115, 120, 193, 247
Brjarne Wong 282
Brother of Darkness 202, 211, 223, 247
Buddhism 127
Bu Wancang 283
Cai Chusheng, 75, 251
Call If You Need Me 188, 248
Category III Hong Kong horror films 21
Chen Kaige 39, 143, 156, 186, 254
Chi Lee 249
Chinese dialects 10, 182-183, 185
Chinese family 126, 128-129, 131-132, 192
Chineseness 15-16, 27, 42, 46, 200, 260
Chinese-ness 21, 181-182, 186-187
Chiu Keng Guan 185, 249, 253
Chodorow 103-104, 110, 261
Christopher Sun Lap Key 247
Chui Mui Tan 253
Chung Siu-hung 250
Clara Law 209, 249
Clarence Ford 214, 251
closet 19, 35, 113-114, 116, 119-125, 129-132
Close-up 11, 24, 55, 58, 62, 74, 77, 84, 142-143
clothing 74, 76, 83, 139, 142-143, 147, 152
coming out 5, 113-115, 119, 123-129, 278
Company of Mushrooms 184, 248
Confucian 72, 104, 106-107, 126-127, 202, 211, 214, 231
consumption 22, 72-73, 143, 149, 190, 193, 195, 225, 227-228, 234
Corey Yuen 252
corporeal 9, 13, 19, 46, 59, 64, 71, 76, 89, 92, 98, 118, 128, 131, 163-165, 170, 173, 193, 213, 263
Cosmetic 231

cosmopolitan 72-73, 80-81, 83, 86, 88, 90, 94, 96, 98, 184, 265
costume 43, 82, 138, 149, 153-154, 197, 223
Crazy 168, 216, 242, 248
Creolization 98, 109, 267
cross-dressing 18, 83, 95, 100-101, 106, 110, 219, 257
Dai Jinhua 69-70, 79, 85, 87-88, 91
Danny Lee 248, 253
Daughter of Darkness 202, 211, 248
Daughter of Darkness 2 211, 248
Deleuze 64, 67, 140-141, 144-146, 150, 156-157, 160, 170, 179-180, 258, 262, 268-269, 271, 278
Dennis Wu 247
desire 1-3, 7, 9, 11, 14-15, 17-18, 21, 23, 29, 31, 36, 43-46, 51, 54-55, 57-58, 61, 64, 69-74, 78-79, 82, 84-85, 87-88, 92, 102, 110, 115, 117, 119, 121, 124, 126-127, 131, 155, 169-172, 180-183, 188-189, 195-198, 203-204, 206, 208, 212, 214, 230, 241, 245, 257
diasporic 16-17, 27, 36, 42, 46-47, 89, 94, 99, 182, 263
Disposability 7, 159
documentary 10, 19-20, 137-139, 141, 143, 147, 154, 159-160, 165, 169-172, 175-176, 178-180, 259, 261, 268, 271
Don't Play with Fire 208, 248
Dreaming the Reality 215, 248
East is Red, The 219, 248
East Palace, West Palace 17, 28, 41-42, 248
eating 6, 21, 97, 114, 128, 148, 181-182, 188-197
Ebola Syndrome 209, 220, 223, 248, 275
Edward Lam 123-124, 134
Elephant and the Sea 191, 248
Elisabeth Grosz 71
embodiment 5-20, 22-24, 26, 28, 34-35, 40, 42-43, 51, 54, 58-59, 61, 67, 69-76, 81-85, 88, 114, 137, 139-141, 148, 151-153, 159-161, 163-166, 170, 225, 227, 230, 232-234, 238, 241, 243-245, 255, 269, 273-274, 277
Enter the Dragon 17, 28, 35, 249
Eric Khoo 98, 247
Eros 5, 17, 51-59, 61, 64-65, 204, 206, 210, 218, 222, 224, 249, 268
eroticism 22, 57, 61-62, 225, 231, 239, 242, 244

Eternal Evil of Asia, The 213, 249
ethics 1-3, 7, 9-10, 14-17, 27-28, 31-33, 35, 41-43, 46-47, 103, 126, 177, 221, 277
exploitation 21-22, 175, 201-202, 207, 222
Face 127-128, 249
failure 1-3, 7, 9, 15, 20-22, 25, 31-34, 40-41, 45-46, 65-66, 69-74, 79-81, 86-88, 106, 142, 144-145, 149, 151-152, 159-161, 163, 165, 170-171, 173, 177, 199, 206, 210, 217, 221, 225, 227, 230, 235, 238, 244, 255, 264
Farewell, China 209, 249
Farewell My Concubine 17, 28, 39-41
female stars 18, 69-71, 73, 80, 91-92
female subjectivity 18, 69-70
feminine pornography 22, 244
Feminism, 7, 91, 110, 225, 241
FINAS 184-185, 199-200
*Flirting Schola*r 209, 249
Flower in the Pocket 182, 185, 193, 249
Fuck Cinema 169-170, 176, 249
gay 10, 42, 111, 124-125, 127, 129, 131-133, 216, 265
gaze 12, 18, 22, 31-32, 34, 36-37, 46, 48, 69-70, 74, 78-79, 87, 117-118, 121, 131, 172, 174, 213, 216, 225, 239, 244-245
Gerald Lee 247
Give Me a Home 117-118, 249
Goddess 75-79, 85, 89, 93-94, 249, 264
Godfrey Ho 253
Grandmasters, The 249
Great Day 185, 249
Happy Times 17, 28, 38, 41, 249
He and She 209, 249
He's a Woman, She's a Man 209, 249
Herbert Marcuse 21, 201, 222, 224, 268
Her Fatal Ways 208, 250
Herman Yau 209, 248, 253
Hero Dream 208, 219-220, 224, 250
Heroic Trio, The 250
Her Vengeance 213, 250
heteronormative 22, 119, 122-124, 126, 201, 212-213, 216, 220, 235

heterosexual 104, 114, 121, 124, 214-215, 219, 228-229, 234
Hoi Mang 247
Hokkien dialect 101
Homecoming 102, 111, 187, 250
homophobia 34, 128, 215
homosexuality 39, 49, 106, 111, 118-119, 122-124, 126, 155, 265, 267
Hong Kong 6, 17, 21-22, 27, 36, 46, 51-52, 67, 69, 73, 76, 80, 83, 86-87, 90-91, 94, 123-124, 128, 134-135, 139, 148, 154, 178-179, 185-187, 194-195, 201-202, 205, 207-219, 222-232, 238, 240-243, 245-255, 257, 259, 266, 268, 271, 275
horror 6, 21-22, 168, 171, 185, 201-202, 207-213, 218, 223, 231
Hou Hsiao-Hsien 43, 253
I Don't Want to Sleep Alone 117-118, 134, 155, 194, 250, 259
Ho Yuhang 182-183, 186, 191, 194, 200, 247, 252
Ice Kacang Puppy Love 182, 184-185, 187, 250
Ien Ang 98, 106, 109
If It's Not Now, Then When? 184, 250
Illegal Immigrant, The 250
In the Mood for Love 17, 27-29, 31, 43, 46, 48, 62, 89, 93, 250, 260, 266
inutile 20, 137, 141-142, 144-150
Jackie Chan 36, 251
Jack Neo 98, 101-103, 107-108, 251, 254
James Lee 182-183, 191, 194, 247-248, 250-252
Jia Zhangke 19, 137-138, 140, 143, 150, 152, 155-157, 160, 175-176, 253-254, 258, 268, 273, 275
John Hau 223, 248, 251
Johnny Lee 248
Jonathan Kaplan 247
Judgment Day 96, 108, 250
Judith Butler 115, 133, 259
Ju Dou 17, 28, 36, 250
Julian Lee 216, 247
Julie Lee 217, 224, 253
Karamay 6, 20-21, 159-178, 250
Khoo Eng Yow 182-183
Kinta 1881 185, 250
Kirk Wong 250

Kuo Chu Huang 251
Lai Kai-Ming 248
Last Temptation of Christ, The 250
Lawrence Cheng 249
Lee Kang-sheng 113-114, 133-134, 136, 266
Lee Thean-Jeen 250
LGBT 100, 217
Liew Seng Tat 183, 249
Lik-Chi Lee 249
Love Conquers All 185, 188, 241, 250
Love to Kill 211, 250
Love Without End 80, 85, 94, 250
Lung Sang 252
Lust, Caution 17, 28, 43, 251
Lu Xun 126, 135, 268
Mabel Cheung 250
Malay 21, 97-98, 100, 108-109, 181, 183-184, 188, 190-191, 193, 197, 200, 273
Malaysian cinema 6, 21, 181-185, 188-190, 195-197, 199-200, 267
Malaysian Sinophone cinema 21, 181-184, 187-188, 193-195, 197
Man Kei Chin, 249
Mao Kurata 254
Marcuse 21-22, 201-207, 210, 212-214, 218, 220-222, 224, 268
Marxism 206
masquerade 80-82, 84-86, 88
matrophobia 106
Merleau-Ponty, 69, 90, 92, 265
Michael Chuah 185, 252
mirror 5, 19, 113-116, 119-123, 127-128, 130-132, 135, 191, 262
modes of address 51, 54-57, 60-61, 63, 65-66
mothering 18, 95, 103, 105-107, 110-111, 259, 261
mothering, 95, 105
movement, 59, 61-62, 90, 93, 125, 138-139, 141, 146, 157, 269-270
Mr. Vampire 219, 251
multi-ethnic 197
multilingual 96-97, 108, 184, 273
Naked Killer 214, 251

Nasi Lemak 2.0 185, 251
national allegory 98
national identity 6, 21, 52, 67, 181-182, 185, 188, 198, 200, 245, 266, 272
New Documentary, 159
New Media, 109, 225, 265
New Woman 78, 90, 93-94, 264
New Women 75, 77-79, 85, 87, 251
Ngai Choi Lam, 250
Nietzsche 69-70
nudity 56, 207, 216, 219
Offense Storm 215, 251
Once Upon a Time in China IV 209, 251
Ong Kuo Sin 108, 250
Pantyhose Heroes 208, 251
Passionate Killing in the Dream 215, 251
Passion Unbounded 216, 223, 251
peranakan 97-98, 100-101
performativity 19, 42-43, 81, 104, 113, 115-116, 132
Petaling Street Warriors 187, 251
Peter Chan 249
petition 167, 173, 176, 251
phronetic 21, 159-160, 176
pleasure 12, 18, 34, 69-70, 81-82, 86, 127, 131, 203, 206-207, 210, 218, 226-227, 233, 235-236, 238, 241, 255, 274
Police Story 36, 222, 251
pornography, 109, 225-226, 228, 231-233, 238
Possessed 185, 251
Post-Cinema, 225
psychoanalysis 55, 110, 202-203, 212, 261, 279
qi 72, 74
qipao 33, 76-77, 83
queer 7, 10, 19, 34, 40-41, 106, 110-111, 113-116, 118-126, 128, 130-133, 135-136, 155, 208, 214, 216, 220, 225, 228-231, 234-235, 241, 243-245, 255, 260, 264, 267-268, 271, 273, 277-279
queer cinema 41, 106, 110, 123, 155, 268
Queer Studies 225

Rain Dogs 194-195, 197, 252
Raise the Red Lantern 36-37, 49, 252, 273
Raper 213, 248
Raymond Lee 248
Rebels of the Neon God 114-115, 117, 121, 252
Red Detachment of Women 167, 179-180, 252, 261
Richard Dyer 73, 92
Ricky Lau 251
ritual 159, 165, 167
Robert Clouse 249
Rock on Fire 215, 252
Room to Let 189-190, 252
Royston Tan 98
Ruan Lingyu 5, 17, 69-70, 75, 89-90, 93, 264
ruined bodies 5, 16-17, 27-28, 32-34, 36, 41-43, 46-47
Run and Kill 211, 252
Sammo Hung Kam-bo 251
Sanctuary 183, 186, 194, 252
Seed of Darkness 185, 252
Sensuous 12, 26, 51, 53, 58-59, 63-64, 67, 166, 269
Sepet 192, 252
Sex and Zen 22, 207, 225, 227, 231, 235, 240, 245, 247, 252
sexuality 10, 49, 56, 69, 76, 87, 91-92, 111, 114, 120-121, 124, 128, 130, 202, 215-216, 219, 222, 226, 234, 241, 245-246, 255, 261, 264-266, 271, 275, 277-278
Singapore 5, 7, 18, 95-102, 106, 108-111, 185, 187, 194, 199, 247, 250-254, 257, 260, 264-267, 272-274, 277-279
Siniticate 5, 18, 95-98, 101, 107
Sinophone 6-7, 9, 11, 21, 96-98, 108-109, 165, 181-195, 197, 272-273, 278-279
Sinophone films 21, 181, 184-185, 187, 189-191, 194
Sinophone languages 21
Soderbergh 52, 55, 249
South of South 192, 252
spectatorship, 51, 85, 87, 245, 279
Spider Lilies 17, 28, 41, 253
Spinoza, 141
Still Life 155, 176, 253, 259

Story of Qiu Jiu, The 253
Stray Dogs 114, 117, 253
subaltern 96, 116-117, 119, 122, 125
Sweet Smell of Death, The 253
Tan Chui Mui 182-184, 248, 250, 252
Tang Huang 248
Tao Qin 250
temporality 6, 137-141, 143, 147-149, 151-152, 155, 157, 174, 258
The Beautiful Washing Machine 182, 190-191
The Closet in the Room 119
The Goddess 75-79, 85, 89, 93, 264
The Hand 5, 17, 51-54, 56-62, 64-67, 249, 274
The River 123, 126-128
Three Times 17, 28, 43, 253
Tiger Woohoo 185, 253
togetherness-in-difference 95-96, 98-99, 106, 109
tongzhi 114, 129, 133, 135-136, 261, 265, 274-275
trans-Asian 6, 22, 225, 228, 230
Transgender 99-100, 107, 109, 219, 265, 267
transgressive 206-207, 219
trans-mothering 5, 18, 95, 97, 99-100, 103, 106-107
transnational 1-3, 5, 7-11, 15-18, 21, 23, 25, 27-29, 31-32, 34-36, 41-42, 45-48, 51-54, 57-58, 60, 62, 65-66, 80, 90, 93, 107, 110, 114, 138, 181-182, 185-187, 197, 200, 228, 258, 260, 264, 267, 279
transnational cinema, 51, 53
Trilogy of Lust 217, 224, 253
Tropical Malady 23-25
Twist 127, 211-213, 253
Uncle Boonmee Who Can Recall His Past Lives 17, 28, 45, 253
Under Police Protection 215, 253
Untold Story, The 253
Useless 143-145
utile 142, 144-150, 152
violation 202, 208
visibility 12, 18, 69-70, 72-74, 79, 82, 85, 88, 100, 178, 226, 261
Vive L'Amour 117, 121, 253

Walker 6, 19, 105, 110-111, 137-142, 147-149, 152, 154-155, 253, 270, 274
We Not Naughty 187, 254
What Time Is It There? 128, 254
Wilson Yip 248, 253
Women at 40 97, 101, 108, 251
Wong Chun-Yeung 248
Wong Kar-wai 5, 17, 27-29, 43, 46, 48, 51, 56, 76-77, 89, 93, 186, 247, 249-250, 259, 266
Wong Ying-Git 251
Woo Ming Jin 248
Wuniu 209, 254
World, The 254
Wu Yonggang 249
Xie Jin 252
Xinjiang 159, 161-162, 167, 178
Xu Xin 6, 20, 159, 161-167, 169, 171-178, 180, 250, 269
Yellow Earth 143, 156, 254
Zero Chou 253
Zhang Yimou 36, 249-250, 252-253
Zhang Yuan 248